GIS Online

INFORMATION RETRIEVAL, MAPPING, AND THE INTERNET

Brandon Plewe

OnWord Press
Thomson Learning™

Africa • Australia • Canada • Denmark • Japan • Mexico • New Zealand
Phillipines • Puerto Rico • Singapore • United Kingdom • United States

OnWord Press Staff

Publisher: Alar Elken

Executive Editor: Sandy Clark

Managing Editor: Carol Leyba

Development Editor: Daril Bentley

Editorial Assistant: Allyson Powell

Executive Marketing Manager: Maura Theriault

Executive Production Manager: Mary Ellen Black

Production and Art & Design Coordinator: Leyba Associates

Manufacturing Director: Andrew Crouth

Technology Project Manager: Tom Smith

Cover Design by Lynne Egensteiner

Library of Congress Cataloging-in-Publication Data

Plewe, Brandon, 1968-
    GIS online : information retrieval, mapping, and the Internet /
Brandon Plewe.
        p. cm.
    Includes index.
    ISBN 1-56690-137-5
    1. Geographic information systems. 2. Internet (Computer network)
I.   Title.
G70.212.P57 1997
910'.285—dc21   97-22624
                                                        CIP

For more information, contact

OnWord Press   An imprint of Thomson Learning

Box 15-015   Albany, New York USA 12212-15015

About the Author

Brandon Plewe is an assistant professor of geography at Brigham Young University in Provo, Utah, specializing in GIS and cartography. While a graduate student at the State University of New York at Buffalo, he was one of the pioneers of distributed geographic information (DGI)—the integration of GIS and maps with the Internet—starting Virtual Tourist and developing the TIGER Mapping Service with the U.S. Census Bureau. He also privately practices as a cartographer and DGI consultant.

Acknowledgments

Many people have contributed to this book. I wish to thank the following people, who discussed various issues with me, and helped get many of the facts straight: Kurt Buehler, OGC; Barbara Buttenfield, University of Colorado; Allan Doyle, BBN; Benoit Galarneau, LAS; Simon Greenman, Geosystems Global; Susan Huse, University of California-Berkeley; Clinton Libbey, Core Software; Mike McGill, Autodesk; Doug Nebert, FGDC; Ian Nixon, Intergraph; See Hean Quek, Universal Systems; Chris Stuber, U.S. Census Bureau; Berry Taylor, Bentley.

The screenshots and Web pages used in this book are courtesy of the following organizations: Ameritech, Autodesk, BigBook, City of Oakland, Core Software, Etak, Genasys, Geosystems Global, Intergraph, MapInfo, Object/FX, Safe Software, Solid Systems CAD, University of California, University of Minnesota, Universal Systems, Vicinity, Visa, and Yahoo.

Much of the research for this book was supported by a grant from the BYU College of Family, Home, and Social Sciences. I would also like to thank Kalvan Hone, who did considerable work evaluating some of the software, and Eric Ringger for putting me on the winding road many years ago that eventually led to this book. I thank my wife for putting up with a husband who is far too busy than is probably good for him.

Brandon Plewe

Contents

Chapter 3 A Host of Solutions 63

Chapter 4 Commercial DGI Software 97

Introduction

Purpose of the Book

This book is designed to help you participate in the historic development of the Internet. It will guide you through the development of distributed geographic information (DGI) applications, which allow you to use the Internet or your intranet to distribute geographic information to a broad audience. A wide variety of DGI services have already been created by many organizations, enabling either selected individuals or the general public to use maps, GIS data, and other forms of information.

This new technology can therefore be used to further the goals of your organization, be it yourself, a business, a government agency, an educational institution, or any other group. These goals might include education, providing better service to customers and clients, and profit.

Unlike many other Internet books being published, this is not just a directory of interesting DGI sites. There are a large number of services out there, but sites on the Internet change so rapidly it is really not possible to create such a listing and expect it to stay current for more than a month.

Although reference is made to many sites herein, these are only a small sample of the total available. They were selected because they are exemplary of various aspects of DGI, and because they are relatively stable. A more complete and up-to-date listing is given at the Web site that accompanies this book, at *http://kayenta.geog.byu.edu/gisonline.*

Neither is this book a technical manual or step-by-step guide to building a DGI site. The wide variety of possible services and the fundamental differences between available software platforms make that type of book nearly impossible when trying to cover the scope of DGI. It is expected that the reference manual for the software you choose to implement will give you that detailed technical information.

This book will, however, give you a perspective on the subject of GIS online that is much more broad than a perspective gained by looking at one or two existing sites or talking to a single GIS vendor. It is hoped that you will be able to use this book to leverage your geographic information in the most effective way possible—by creating online services that are powerful, professional, simple, and efficient.

The information contained herein will be useful to you in every phase of the development of your geographic services. This includes planning the service, implementing it on the computer, and maintaining it over the long term.

Audience

This book is intended for both the technician and the decision maker. To use this book effectively, you do not need a great deal of technical knowledge, but you should at least be somewhat familiar with the general concepts of both GIS and the Internet.

↝ NOTE: *If you are unfamiliar with the subjects of GIS and the Internet, you should review the glossary of Internet terms, especially the technical aspects of Web servers.*

Structure and Content

Creating a DGI service can be a long, involved process, from the early planning and design of a site to its implementation and long-term maintenance. Therefore, this book is organized according to this process, guiding you through each phase.

This book covers a broad spectrum of applications you may wish to implement in your organization. DGI services can be valuable for multinational corporations, small businesses, nonprofit organizations, government agencies, academic researchers, and even individuals. This book covers this technology as it applies to all of these entities.

Chapter Content

The book begins, in Chapter 1, with an introduction to the basic concepts of how maps and geographic information can be distributed using Internet technology. A brief history of the development of this field is included, as well as an indication of the varied services developed to date.

The majority of the book describes the process of planning and designing your own DGI service. Chapter 2 will help you design a general strategy for your services: what information you wish to make available to whom, and at what price. Chapter 3 contains details about the types of services you can provide, inclding dynamic maps, spatial queries, full GIS analysis capabilities, and distribution of raw data to users with their own GIS software.

Many of the early prototype sites were custom built from the ground up. However, this approach is not for everyone. Chapter 4 looks at many recent commercial products and helps you decide which ones might work for your particular applications.

Before you can create your services and build your site, you need to come up with the necessary labor and capital resources. Chapter 5 describes these resources.

Chapter 6 discusses the details of the design for the appearance of your site. The chapter contains guidelines for the design of maps, as well as for the various functions of your DGI applications. It also contains information on how these graphic elements will interface with your DGI program.

Chapter 7 guides you through the process of physical implementation, including programming, data preparation, announcing your finished service to potential users, and permanent maintenance. Chapter 8 looks at several (potentially difficult) issues you will need to consider throughout this process (as well as in the long term), such as copyright protection, privacy, and bias.

Chapter 8 also looks ahead to technologies and institutions you can expect to see in the "next generation" of DGI. It follows in detail the creation of a suite of services built in a real application, the NSDI node for a state government.

The Glossary

This book focuses on the creation of GIS-enabled Internet services. It does not attempt to discuss all of the technical details of GIS or the Internet. It is assumed that the reader has at least a basic knowledge of the concepts of GIS and the Internet. However, because of the wide variety of organizations that may be interested in using the techniques discussed herein, an illustrated glossary is included.

Some readers may not have an extensive GIS background, whereas others may be GIS professionals who have not built Web servers. Numerous entries in the glossary deal with the technical aspects of the Web that are of particular importance to those who run servers.

The glossary is included as an introductory resource to the language and many of the concepts of GIS and the Internet, focusing on those terms that are used frequently in the book. The entries in the glossary are intended as brief, descriptive definitions, not complete technical explanations. For those who need more information, there are many excellent books that provide a more in-depth introduction to either the Internet or GIS.

1

Distributed Geographic Information

Spatial Technology for Everyone

Although it has been around for 30 years, only since 1995 has the Internet emerged as a potentially dominant force in global communications. We are only beginning to catch a glimpse of the long-term ramifications of this "network of networks," but it is already becoming part of the infrastructure of cultures around the world.

In some respects, the Internet has the elements of a temporary fad. Its extremely rapid growth, general lack of serious content, and large numbers of casual users or curiosity seekers combined with recent warnings that the infrastructure may not be able to survive its own popularity suggest that it may not last forever. However, amid the fluff there are many areas in which the Internet is extremely valuable.

In fact, the Internet has already become an integral part of much of society. This system has revolutionized journalism (due to online up-to-the-minute news), science (because of the capacity for global cooperative research), publishing (because of the ease with which anyone can publish their ideas), and many other fields. These activities deal with and disseminate truly powerful content. This is where the Internet has its real and lasting value, and why it will continue to grow, develop, and become even more ensconced in society.

Geographic information systems (GIS) has the potential to be one of these fields. The Internet holds promise for exponential increases in the efficiency and effectiveness of the ways in which we obtain, use, and share geographic information in all its forms (including maps, graphics, text, and data). Many extraordinary systems have already been built, and over the next few years, an increasing number of GIS applications will "go online."

What Is DGI?

A broad array of products and services are discussed in this book. These products and services deal with the use of Internet technologies to give people access to geographic information in a variety of forms, including maps, images, data sets, analysis operations, and reports.

Distributed geographic information (DGI) is a term proposed to refer to this entire field; that is, the widespread (i.e., to a larger audience than would have access using traditional GIS technology) distribution of geographic information in any of the forms previously mentioned.

DGI applications range from simple, pre-drawn maps on a Web page to network-based collaborative GIS in which GIS users at remote locations share common data and communicate with one another in real time (not yet widely available). The technologies being developed to make DGI applications possible include servers (which store data and applications), clients (which use the data and applications), and network communications (which control the flow of information between servers and clients).

On the server (dissemination) side of the equation, issues include such things as the speed of query and transfer, the possibility of including full GIS functionality or just basic mapping, and the efficient storage of and access to large quantities of geographic information. On the retrieval side, issues include effective user access to and location of desired information, as well as means of viewing and analyzing the information, often with GIS software unfamiliar to the user.

Although there has been some basic research in this field, most of the development to date has been through the experimental development of both prototype and production services by a wide variety of organizations. There is a growing DGI industry as well, with many GIS vendors and third-party companies developing software that enables the widespread distribution and retrieval of geographic data, especially over the Internet. Although most of the software is relatively new, this book offers an overview of most of the available DGI software packages (see Chapter 4), each of which has its own strategy for delivering geographic information.

The primary goal of this book is to help you use the principles of DGI to develop one or more applications by which you can distribute geographic information—whether it be GIS data, graphic maps, or even textual reports—as part of the mission of your organization. Although most readers will be using an existing GIS as the source of their data, this is not necessary, as many DGI systems have been built on pre-packaged data for non-geographical organizations.

The Basic DGI Service

There is a wide variety of ways in which you can distribute geographic information on the Internet, but they are all founded on the same general design. The basic architecture is similar to the client/server model shown in the following illustration—a model upon which the World Wide Web (WWW; hereafter referred to as the Web) and most other Internet services are based.

The classic model includes a client program (a Web browser such as Netscape), which makes a request to a server program [i.e., a specific page by its uniform resource locator (URL) address]. The server processes the request and returns the information [e.g., a hypertext markup language (HTML) page] to the client.

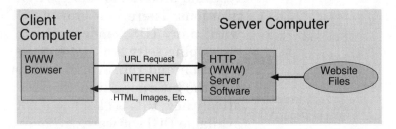

Architecture of the World Wide Web, a classic client/server design.

The model used in DGI is an extension of the client/server concept, known as a multi-tiered server. A client is typically a Web browser or other Internet access software such as the File Transfer Protocol (FTP). It is even possible to conceive of an entire GIS program as a DGI client. The multi-tiered server consists of a normal Web server and a running GIS program, with communication between the two facilitated by a specialized DGI program.

When someone makes a request for a map or other GIS product, the request message is sent over the Internet to the Web server. The server recognizes it as a DGI request, and passes it to the DGI program, which translates it into some type of internal code (e.g., queries or map drawing commands). This is passed to the GIS software, which processes the request, usually using custom *scripts* (programs interpreted by the GIS software).

The software then returns the result, which could be a map, text, or even a raw data file. The DGI software is then responsible for reformatting the output into an Internet-standard format, or at least something understood by a browser plug-in or Java applet. (An applet is a module that extends the functionality of a Web browser.) The reformatted information is then returned by the Web server to the client, where it is displayed. This display might in turn be used to issue a new request, such as if someone were to click on the map. This pattern repeats itself many times during a DGI session.

The multi-tiered client/ server model of most DGI services.

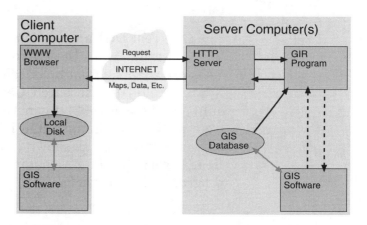

In many cases, the architecture of a site may differ from this general model. Quite often, the DGI software incorporates data reading, mapping, and basic analysis capabilities, which eliminates the need for dedicated GIS software. This is commonly found in simpler applications such as street map browsers, which come with their own pre-packaged data.

A History of DGI: The Early Years

To get an idea of what DGI services you might incorporate into your own organization, you should have an idea of what others are doing. There is a wide variety of applications already running that are very useful in accessing geographic information. This field has grown very rapidly (in what other context could you call 1993 "early"?), with just a five-year history, but many of the events during this short period are instructive.

The First Prototypes: Xerox and Virtual Tourist

The rapid development of DGI has closely followed the rapid development of the Web. As the Web started to gain popularity in early 1993 (it was originally started in 1989, but took a while to catch on in the Internet community), some saw maps as a valuable resource to have online.

The first real DGI application (although not referred to as such), was the Map Server (*http://map web.parc.xerox.com/map*), developed by Steve Putz at Xerox's Palo Alto Research Center (PARC) and put online in June of 1993. It generated very simple maps from public domain data, and included an interactive browsing interface, as shown in the following illustration. It was based on a custom-built map-generating program (Map Server) and used no commercial GIS.

A sample of the Xerox PARC Map Server, the first DGI application.

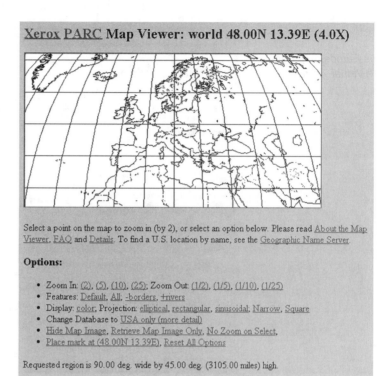

This first application helped spawn other ideas for incorporating maps into the Web. In November of 1993, staff at Tromsø University in Norway created a map highlighting the Web servers in their country (*http://www.uit.no/norge/homepage no.shtml*).

This idea quickly caught on in other countries because having a large number of Web sites was a source of national pride for many of the Web's pioneers. Within a few months, maps for many countries were online, and Virtual Tourist (VT; *http://www.vtourist.com/web map*) was created in January of 1994 as a map-based index to these sites. Today, VT continues as the largest geographical index to the Web. A VT map of Europe is shown in the following illustration.

The Europe map from the Virtual Tourist.

the Virtual Tourist

Europe

The following map leads you to maps and lists of WWW servers that are available in each country. This **does not** contain information *about* countries: that information is contained in City Net.

Select a country from the map below:

With the recent addition of Bosnia & Herzegovina, there are now web servers in every country in Europe. Congratulations!

Both the Xerox Map Server and VT have been very popular over the years, and maps from these sites have found their way into many Web sites (some legally, some not). For example, international and global businesses use continental maps from VT to display the locations of their operations. Environmental activists use these same types of maps to display areas in which they are interested, and even organizations such as bike clubs use them—to chart the locations of trips. This varied and widespread popularity has verified the idea that spatial information is an extremely useful part of many Web applications.

DGI and NSDI, 1994

In 1994, a large number of prototype projects for distributing spatial data on the Internet were introduced. They came from such places as government agencies in the United States, Canada, and Australia; universities in the United Kingdom and the United States; and even a few private companies. Most either used predesigned raster maps (placed in HTML pages the same way any other images are) or a true GIS to generate the maps.

The latter solution was very slow because the GIS software was not made to handle live connections in this manner, and had to be restarted to process each map request. The researchers building these applications began to come together to discuss common issues, and DGI as a research field and industry had begun.

The other major DGI development of 1994 was the initiation of the National Spatial Data Infrastructure (NSDI; *www.fgdc.gov*)—the home page for which is shown in the following illustration. This U.S. federal initiative was begun with a presidential executive order (No. 12906), and has been run by the Federal Geographic Data Committee (FGDC), a task force representing all federal agencies whose mission involves the production or use of geographic information. Most of these agencies fall under the departments of Defense, Interior, Commerce, and Agriculture.

*Home page
of the NSDI.*

Federal Geographic Data Committee (FGDC)

The National Spatial Data Infrastructure (NSDI) encompasses policies, standards, and procedures for organizations to cooperatively produce and share geospatial data. The Federal Geographic Data Committee(FGDC) has assumed leadership in the evolution of the NSDI in cooperation with state and local governments, academia, and the private sector.

FGDC Organization

Web Calendar of Events

New Funding Opportunity: NSDI Benefits Program

The National Geospatial Data Clearinghouse

Geospatial Data query form

Other Geospatial Data Sources

[Federal][State][University][Foreign][Commercial][Miscellaneous]

NSDI laid the groundwork for the effective sharing of data by federal, state, local, and private organizations with each other and with the general public. NSDI's work consists of training programs, standards development, project grants, federal mandates, and a network of contacts for coordination of data production efforts. All of these tasks have had a major influence on DGI, and on GIS in general.

All U.S. federal agencies with geographic information have begun to put the information online. FGDC has been involved in educational and cooperative efforts that have aided many state, local, educational, private, and even international GIS producers begin doing likewise.

Another important development of late 1994 was the funding of the Digital Libraries Initiative by the U.S. National Science Foundation. Among five projects funded was the Alexandria Project (*http://alexandria.sdc.ucsb.edu*), housed at the University of California at Santa Barbara (UCSB). This project aims to build an online digital library of spatially referenced information by 1998. This library will allow people of various backgrounds to locate, view, and analyze digital spatial information (the goal is to have a million holdings by 1998) on the Internet.

Unlike many other prototypes and production services being built and discussed herein, this project is focusing on basic DGI scientific research. Scientists at UCSB and other locations are investigating issues such as the user interface, the search mechanism, and the efficient storage and retrieval of very large data sets. Their research promises to add considerably to the development of most future DGI services.

TIGER Mapping Service and Its Successors, 1995

The main innovation of 1995 was the development of live mapping engines, building on the concept behind Xerox's original server. One of these is the TIGER Mapping Service (TMS; *http://tiger.census.gov*), developed by the U.S. Census Bureau. It is a prototype system for delivering fair-quality street maps (from TIGER data) interactively. It also produces simple thematic maps from 1990 Census statistical data. The following illustration shows a TIGER Mapping Service map of median income.

Click ON THE IMAGE to:

○ Zoom in, factor: [2]

○ Zoom out, factor: [2]

● Move to new center

○ Place Marker (select symbol below)

○ Download GIF image

OR

[REDRAW MAP]

with any option selected below

OFF/ON Layers
- ☐ ☐ City labels
- ☐ ☐ Grid (lat/lon)
- ☐ ☐ Cens bg points
- ☐ ☐ Cens bg bounds
- ☐ ☐ Congress dist
- ☐ ☐ Counties
- ☐ ☐ Indian Resv
- ☐ ☐ Highways
- ☐ ☐ Parks and Other
- ☐ ☐ MSA/CMSA
- ☐ ☐ Cities/Towns
- ☐ ☐ Railroad
- ☐ ☐ Shoreline
- ☐ ☐ Streets
- ☐ ☐ Census Tracts

OFF/ON Layers
- ☐ ☐ Interstate labels
- ☐ ☐ St Hwy labels
- ☐ ☐ State Bounds
- ☐ ☑ US Hwy labels
- ☐ ☐ Water bodies
- ☐ ☐ Zipcode points

Scale: 1:212162 (Centered at Lat: 40.73889 Lon: -111.89349)

[REDRAW MAP]

If your browser doesn't support client-side imagemaps, use the controls below to navigate the map.

	NW	N	NE	
ZoomIn	W	Pan	E	ZoomOut
	SW	S	SE	

Map of median income from the TIGER Mapping Service of the U.S. Census Bureau.

As with the Xerox server, TMS uses a custom map generating program rather than a commercial GIS to deliver maps quickly. Although the service does not have all of the functionality one would like to see, its popularity and usefulness for a large audience has engendered two types of DGI services.

One group includes several professional "online road atlas" sites introduced over the past couple of years from private companies with their own U.S. or even global databases. These sites include the following.

- MapQuest (*http://www.mapquest.com*) by Geosystems Global

- MapBlast (*http://www.mapblast.com*) by Vicinity

- EtakGuide (*http://www.etakguide.com*) by Etak

- GridNorth (*http://www.gridnorth.com*) by Autodesk

Related sites include thematic mapping engines for producing maps from census and other statistical data. Notable examples include the Demographic Data Viewer (*http://sedac.ciesin.org/plue/ddviewer*) by CIESIN and the Census Bureau's own Data Access and Dissemination System (DADS), currently under development. This latter service will build on the thematic capabilities of TMS to give citizens map-based access to a wide variety of bureau data, including not only the decennial Census of Population and Housing but the economic and agricultural censuses and other sources.

Everyone Joins the Party, 1996

The year 1996 might very well go down in the history books as the "Year of the Internet." Whether it lasts in the long term or not, and although still only a small minority of the populace has access to it, the Internet definitely became part of mainstream society during that year. Any computer company without an Internet agenda appeared to be doomed to oblivion.

GIS software vendors were not by any means immune to this wave of online vision. Although third-party companies introduced software for connecting spatial data to the Web at least a year earlier, the technology had not really caught on yet. During 1996, all of the major vendors introduced their long-term Internet agendas, as well as programs for connecting their software to Web servers. This included ESRI, Intergraph, MapInfo, Bentley, Genasys, and others. (DGI software from these and other companies is discussed in detail in Chapter 4.)

Sample Applications

At this time, DGI applications are being offered by a wide variety of organizations. The following are just a few of the most popular sites, which represent very different purposes and approaches to distributing geographic information online.

USGS GeoData Online

`http://edcwww.cr.usgs.gov/doc/edchome/ndcdb/ndcdb.html`

One of the first on the Internet to deliver real GIS data, this site forms part of the U.S. Geological Survey's (USGS) section of the NSDI. As part of its National Mapping Program, this agency produces an immense amount of digital GIS data corresponding to their topographic map series, with themes including transportation, hydrography, terrain, and land use.

The purpose of the GeoData Online site is to allow people to find and download this raw data to use in their own GIS software. Not all of the USGS's data is online yet; there is so much, it just isn't feasible with today's server software and hardware. However, GeoData Online does offer complete small-scale (1:2,000,000) and medium-scale (1:250,000 and 1:100,000) data sets, as well as a small number of large-scale files (1:24,000), which have recently been coming online.

The data files are actually downloaded via FTP, which isn't very easy to navigate, especially with a library as large as this. To aid users in finding the data they need, several interfaces have been provided, including maps (as shown in the following illustration), state-by-state listings, and a large alphabetical list, if you happen to know the name of the quadrangle that covers your study area.

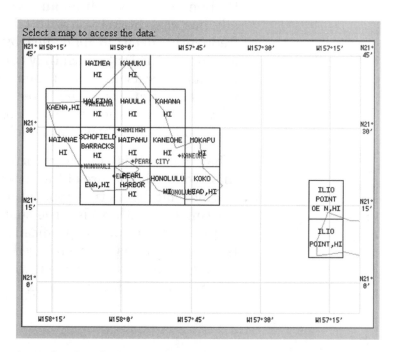

A map interface for locating data sets from the USGS GeoData site.

The GeoData site also provides a complete search engine, which allows you to search the USGS holdings database, which includes information about all of the digital data, both online and off-line. The query can include both spatial and thematic criteria. This useful site has become very popular with data-hungry GIS users (hundreds of files are downloaded daily). However, to the general public without standard GIS software, this data has little value.

VISA ATM Locator

    http://www.visa.com

This very popular site is an excellent example of a DGI-enabled service from an organization that generally has very little to do with GIS. Visa's cardholders frequently need to find nearby automated teller machines (ATM) at which they can get cash advances. This site lets users type in their address, from which the server will locate the three closest Visa-affiliate ATMs, and even show them on a map, along with your site's address (as shown in the following illustration).

What makes this site so important in the context of this book is the nature of the server. The spatial information that drives this application is not based on an existing GIS within Visa; the spatial data and search engine were packaged specifically for the Web site. In fact, the service is not even based on a commercial GIS package, but uses software custom built for this application by a Web design service.

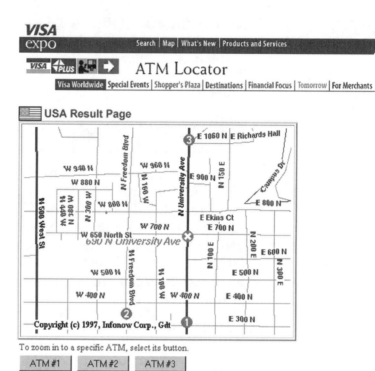

A results page from the Visa ATM Locator, showing three nearby machines.

An organization does not have to be especially GIS-minded to take advantage of spatial information in its Web site. Pre-packaged map data and server software are available, which can be combined with your own data (in the case of Visa, ATM locations). This data can be used to produce a fully functional DGI application, without GIS software.

GRASSLinks

`http://regis.berkeley.edu/grasslinks`

This service, produced by the REGIS program at the University of California at Berkeley, was one of the first examples of a true online GIS, and is still the best example. GRASSLinks provides a wide range of analysis tools—accessible entirely through the Web interface—that allow almost anyone interested in the San Francisco/Sacramento region to study their geographic environment.

REGIS maintains a variety of GIS data sets relating to the urban and natural environments of the San Francisco Bay and Sacramento Delta regions of California. This includes themes such as roads, water features, wetlands, zoning, aerial photographs, and parks. Some of this information is in raster format, and some of it is in vector. Using the REGIS Web site, you can view maps of each of these themes, or display multiple themes together.

This is made possible by a custom-built DGI program that links the REGIS Web server to a running GIS program (specifically, the GRASS GIS software, produced by the U.S. Army Corps of Engineers). Because the GRASS engine is a fully functional GIS, REGIS can do more than just draw maps. The GRASSLinks site can perform many powerful GIS operations on these data sets, including queries, overlays, reclassification, and buffers. (A buffer analysis map is shown in the following illustration.) Even complex, multistep analysis processes can be done. Users can temporarily save the data sets they create on the server and download them for further processing.

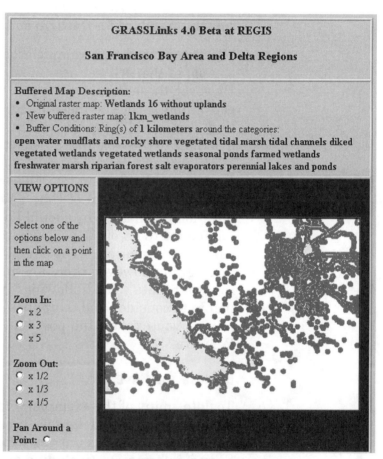

A map resulting from a buffer analysis in GRASSLinks.

The real power of the GRASSLinks site is its accessibility. Normally, only a small group would have access to this information, including the REGIS researchers and local governments, which have their own copies of the data and necessary GIS software. GRASSLinks brings this information to the entire Internet community, which can now perform GIS operations without GIS software. In fact, all you need is a standard Web browser, which makes this a popular site among area residents and businesses, public action groups, and urban and environmental researchers worldwide. It sets the standard for the high end in DGI applications.

The foregoing three examples constitute only a small sample of the DGI services available on the Internet. However, these should give you a good idea of the variety of approaches to distributed geographic information you could take in making your spatial data available to the world (or at least to some portion of it).

Why Should I Share?

Building and maintaining DGI services almost always requires a great deal of time and money. This raises questions such as "Why should I bother?" and "What is forcing me to let other people have access to my hard-earned information?" The truth is, there is nothing forcing you; however, there are many benefits to sharing your data. The following are some compelling points to consider and to apply to your own organization as you explore the possibility of building a DGI Web site.

Because You Have To

To date, many of the exemplary data delivery services have been produced largely because of government mandate. In most countries other than the United States, information (including spatial data) produced by a government agency is owned and copyrighted by the government, which distributes the information according to federal guidelines. Spatial data is often of a very high quality, but can be difficult and expensive for the general public to obtain.

However, in the United States, in principle (more than in practice), the government is considered the property of the people; thus, any information it produces is also owned by the public. This principle, embodied in the Constitution, was codified in the Freedom of Information Act (1965), which states that all information produced by the federal government, except that which impinges on personal privacy or national security, is in the public domain.

U.S. federal agencies must make requested information available to anyone who wants it. They are allowed to charge for this, but cannot profit or even recoup the costs of production. The price of data can only cover the cost of distribution (i.e., the storage medium and the labor of getting the data onto it and shipping it). However, agencies typically have control over the manner in which costs are estimated. Historically, the difficulties and high cost of distribution, as well as a limited supply of software capable of using it, made most spatial data still rather inaccessible to the general public.

With the spread of the Internet, some agencies saw the potential for a relatively painless (and almost costless) distribution of spatial data. These agencies developed the FGDC and the NSDI to leverage the Internet in fulfilling the Freedom of Information Act. In the executive order that started the NSDI, all applicable federal agencies were required to develop a strategy for putting their spatial data online, and to begin carrying out that strategy within six months (early 1995).

Although it is not feasible to put all government information online immediately (the USGS alone has several terabytes of digital data), most of the agencies involved now have sites with libraries of raw GIS data sets ready to download. The USGS GeoData site previously discussed is an example of this type of site. Most of these sites also have associated online services to help users search for the specific files they need, by theme or spatial location.

The Internet has made government data much more available to the GIS community, and some agencies have begun to think of ways to make public information accessible by the general public, who do not have (and likely never will have) their own software. This audience can only be reached by incorporating GIS and mapping capabilities into the Web server so that they can be used with a regular browser. This is the approach taken by the Census Bureau's TMS and DADS, as well as the new National Atlas (*http://www.usgs.gov/atlas*) under development by the USGS.

The Freedom of Information Act and the mandates associated with NSDI do not apply directly to U.S. state and local governments. However, most states now have similar laws and initiatives that require, or at least encourage, them to put data online. States such as Montana and Texas have extensive data distribution sites.

Even though they do not have the same public information foundation, many countries worldwide have seen the benefit of using the Internet as a distribution tool for government spatial data, and have begun initiatives similar to NSDI. A very good example is the National Atlas Information Service (*http://www.nais.ccm.emr.ca*) being put together by Geomatics Canada, the national agency responsible for GIS data, which includes not only raw data but extensive public browsing applications.

Another major initiative is the Australian Spatial Data Infrastructure being developed by the Australia–New

Zealand Land Information Council (ANZLIC, *http://www.auslig.gov.au/pipc/anzlic/anzlicma.htm*), the Intergovernmental Committee on Surveying and Mapping (ICSM, *http://www.auslig.gov.au/pipc/icsm/icsain.htm*), and the Australian Surveying and Land Information Group (AUSLIG, *http://www.auslig.gov.au*).

To Make Money

Spatial information has value. Hundreds of millions of dollars a year are spent to create it or to purchase it, whether digitally or on paper. This is apart from the necessary hardware and software. If your organization produces digital spatial data in a GIS, you have probably spent a great deal of money and time to create it. It certainly has value to you and your clients, but it may be worth something to others as well.

If you have information, whatever it is, chances are somebody out there would like to obtain it. In many cases, a large number of people will be willing to pay a reasonable price for a copy of your spatial data. The Internet has the potential to let them.

Using the available and upcoming technologies for online payment, you have a variety of options for selling your products on the Web. You could create a catalog from which shoppers could buy complete data sets (e.g., Wetland Regions in Cayuga County, New York: $130.). After paying for the information with a credit card, the buyer would be granted privileged access to download the packaged files.

You might charge frequent users a flat monthly fee for unlimited access to your GIS services, whereas non-paying browsers would only have access to some limited capabilities, as a "teaser." Another option would be to let users perform analyses or map requests but ask them to pay for the final results (high-quality maps and reports) using electronic cash.

One of the most difficult issues related to online commerce is determining the appropriate price for your information. This will depend on any costs you are trying to recover, the willingness of your particular audience to pay for valuable information (and on the Internet, they are often not willing to pay at all), and the prices charged by any competitors you may have.

Because You Are Nice

Almost everything on the Internet is freely available. Much of this material is produced by governments, universities, and nonprofit organizations, who are not allowed to make money beyond the cost of producing it. However, a great deal of spatial information on the Internet is contributed by individuals and organizations as a service to the Internet community.

One of the reasons for this, at least in the past few years, is the perspective of those who pioneered the Web and DGI. The value of the Internet in general has always been directly related to the value of its content; the best way to promote the Internet was to add high-value content to it. This was one of the motivations behind the Virtual Tourist and other non-geographic navigation tools: by making it easier to find things, more people could use the net more effectively.

Another motivation is the proof-of-concept idea; that is, many early projects were done just because they were possible. Although the Xerox Map Browser and the TIGER Mapping Service may not have been as powerful or as pretty as the commercial products that followed, they effectively showed that there was a large market for these services.

Because so much good information out there is free, a majority of the online community is not willing to pay for something if they can find a reasonable alternative (even if it is of slightly lesser quality) that costs nothing. You have to work within the expectations of, and alternatives available to, your audience.

You may offer some services free to the public, but providing these services is rarely without cost to you. At the very least you have a Web server to maintain. How can you provide a free service without it being too great a liability for your organization? There are at least two ways to do this. If your site is popular, you could do well displaying banner advertisements, as seen in many sites on the Internet (although making this profitable is more difficult than you might expect). You could also subsidize the maintenance of your free services with the proceeds from your fee services.

To Sell Something Else

Free DGI services are also offered as accessories to higher-value information or products. Sites that offer such services tend to focus on a commercial product or service for which location is important (e.g., store locations, delivery zones, or homes for sale). The geographic service is not the main part of the site, but is used only as a navigational tool to find the product or service applicable to a specific location.

Here, DGI services generally take one of two forms: either maps are used as a query tool (e.g., "Click in the area you would like to live") or as a way of displaying the results of a query (e.g., "Here are the locations of properties for sale in the area you selected"). The maps have little value in and of themselves, but can increase sales of the primary products and services (e.g., property) by allowing users to be more selective and get more detailed information about them.

This is the approach taken with the Visa ATM Locator. They do not expect people to be willing to pay a fee to find the three nearest machines. However, they can reasonably expect that the usage of their ATMs will increase if the service makes them easier to find.

DGI services are sometimes used as part of other DGI services. For example, the USGS GeoData site previously discussed uses an interactive map browser to help users select the exact quadrangle for which they want to download digital data. The maps have little or no intrinsic value but make spatial data at the site considerably easier to find.

To Aid Others in Your Organization

One component of the online world, the intranet, came into its own in 1996. This is a networked system that uses normal Internet applications—such as the Web, e-mail, and teleconferencing—but is designed specifically for personnel within an organization. The Internet community at large is usually not able to access the servers involved. The motivation behind this type of system is that it facilitates the communication of information among everyone in the organization, without the need for people to copy large databases or texts or to install specialized software for each type of information.

Spatial information is one of these special types, which is potentially useful throughout the organization but not easily accessible to the majority of personnel, who are not GIS experts. Using a DGI application, people throughout your organization can access at least some elements of your GIS using nothing more than their Web browsers. Web servers have the ability to control access; therefore, the server can be configured so that certain users have access to certain portions of your information, whereas off-site browsers will have no access to the site.

Intranet DGI applications tend to be much more robust than Internet applications, giving internal users access to a much larger proportion of the data and functionality of the GIS server. Typical types of spatial data shared within an organization include demographics and market research, construction plans, environmental impact analyses, navigation maps and route descriptions, and regional sales and service delivery data.

Summary: One Book, Many Applications

The term DGI, or distributed geographic information, is meant to be inclusive. It is more than putting your GIS online; it includes any application that uses Internet or intranet technologies to make geographic data available to a wider audience than has access to it using traditional GIS software.

The development of applications such as this over the past few years shows the diversity of strategies and techniques that can be employed to make this information available. The three sites described in detail should give some indication of what can be done.

Whereas GeoData Online distributes raw data sets, doing nothing with them and requiring users to have their own GIS software to process them, GRASSLinks includes a wide variety of mapmaking and GIS analysis functions in its service. The processing of the Visa site is somewhere between them, but with an interface designed for the mass market.

From these and other examples, it is obvious that there is not a single solution for delivering spatial information over the Internet. Any DGI service you wish to create will need to be tailored to your situation, and will often be radically different from any existing application you might find in the market.

Thus, this book cannot guide you, command by command, through the process of creating an online application. Rather, it discusses the major steps, helping you decide what you want to accomplish and giving some general directions on how to accomplish it. Before you start, you will need to consider your intended audience, the general goals of your organization, the type of data you have to share, whether your information can be profitable, and which data you want to distribute. The next chapter in particular will help you evaluate your situation and plan a strategy for your DGI applications.

2

Developing an Internet Strategy

The Tedious but Necessary Planning Process

The first step in developing a successful map-based Internet site is the same as that for creating a successful Internet site of any type, as well as that for developing a successful GIS: planning. If you cannot effectively decide what you want to accomplish and how to accomplish it, you will waste precious resources creating applications that do not meet your goals.

Of course, you cannot predict exactly which applications will be successful on your site, or exactly how you will create and maintain those applications. Your plan will change as you begin implementation, and it will change periodically thereafter as you discover which applications are popular and useful, and which ones are not worth maintaining. However, it is better to modify a plan than to work without one. This chapter describes the process of creating a general plan for an Internet site that involves GIS and mapping. Chapters 3 through 6 expand on this topic.

Who Should Be Involved?

Developing a DGI strategy involves answering many questions, and considering many issues from a variety of sources. It is therefore important to draw on the expertise of many individuals. Whether you develop the applications alone or work as a committee, the following experts should be consulted during the planning process. Depending on the size of your organization and the intended priority of your DGI operation, you may choose to include all or some of the following people.

GIS Professionals

In many cases, your DGI applications will be built around an existing GIS. Your GIS staff are most familiar with the GIS software and its capabilities and pitfalls. They may have up-to-date information about what is happening in the DGI industry and what software (see Chapter 4) is available. They are also the people who will likely do the bulk of any DGI application programming required to create an interface between the GIS server and the Web server.

Because of their professional perspective, GIS staff are likely to be enthusiastic about using geography as part of your Web site. They will be able to use this perspective to help determine which goals of your general Internet strategy will best be met using maps and other GIS products.

If your organization does not have a large GIS staff, or does not have a GIS at all (as pointed out by the Visa site example in the previous chapter, one is not always necessary for a successful DGI application), you may not have individuals in this category. However, there will likely be people in your organization who know something about GIS because they are interested in it or use it as part of their work.

Information Systems

GIS, Web sites, and DGI applications are all systems based on computers. Thus, it is important that your computer experts be involved. It is possible that all three programs will reside on the same computer, but more likely there will be at least two machines involved (perhaps more if the popularity of your site requires additional processing capability).

Your information systems (IS) staff will be the most knowledgeable in purchasing the server computer(s), selecting operating systems, installing software, and networking the servers to each other and to the Internet. In most large organizations, the IS department is responsible for housing and maintaining the servers. If your application is part of an intranet site, it is probably your IS professionals who will be managing the rest of that site. Even if IS will not be physically housing your servers, they will probably still be involved in their installation and maintenance.

Webmaster

Because your DGI applications are part of your general Web site (if you have one), the person or people who design, implement, and maintain the general Web site system should be involved in the DGI system. The Webmaster may or may not be part of your IS staff, but will probably be distinct from those involved in the roles described under the previous heading.

Because of the unique and complex nature of DGI applications, it is unlikely that the Webmaster will be heavily involved in their development, at least not as much as the GIS professionals. If you decide to include Java or ActiveX applets as part of your service, your Webmaster will probably be one of the most fluent programmers in these languages.

Your Webmaster will know the most about your Web/Internet server software and hardware, and how other programs (such as DGI modules) can connect to it. Thus, this person will be responsible for keeping the system running smoothly as part of your Web server. He or she is also likely to be an expert in the visual design of effective Web applications, including those with maps.

Decision Makers and Management

In most cases, sharing your GIS capabilities, and your geographic information in general, represents a shift in the guiding paradigm of your organization. Most institutions do not have a history of distributing valuable information to a widespread audience. It is therefore important that those who have a broad vision of the mission of your organization, and are able to alter that mission if necessary, be involved in at least the planning of DGI applications.

Because this group has influence over the allocation of staff and capital, it is important that it share your vision of the importance of distributing geographic information. They must be willing to commit the sometimes extensive resources necessary to create and permanently manage your geographic site.

It is very easy for the technical professionals previously discussed to get caught up in the outstanding capabilities available and the impressive nature of interactive visual applications such as mapping (referred to as the "Kool Factor" in the Internet community) to the exclusion of cost factors and other important considerations. Decision makers have the role of stepping back and looking at proposals from a larger perspective, determining not only what is possible or desirable, but what will be most beneficial and profitable to the organization and its clients.

These people can also be very valuable when the system will involve people outside the GIS/IS staff (the three parties previously described), such as graphic designers and marketing professionals. This varied group of professionals will require a great deal of coordination, and some juggling of time and other resources. Management and its decision makers bear this responsibility and are in the best position to meet it.

Public Affairs/Sales/Marketing/Advertising

Despite the variety of uses and applications you might be designing, a presence on the Internet today is largely a promotional tool. Businesses are using it to advertise products and services to a global market, a fair portion of which has the capability to purchase those products and services online. Even government agencies and nonprofit organizations are dependent on their constituents or clients, and are using the global nature of the network to advertise their goals and programs.

One highly powerful visual promotional tool is the map. Maps are commonly used to sell both ideas (e.g., "Why we are building a mall here?") and products (e.g., "Which Burgerama is closest to you?"), as well as improving the professional image of the organization as a whole. This means that the people in your organization most concerned with its public image will have an interest in creating effective, professional, and visually appealing DGI applications.

Depending on the nature of your organization, you may intend to use the DGI application to sell geographic information, such as graphical maps or raw databases. Apart from an initial investment in getting a site running and supplied with images and databases, many individual maps and databases generally do not have significant direct labor or capital costs associated with them. It can be difficult to determine a fair yet profitable price for information generated largely without labor or capital. You might also use it as a tool for marketing other products, such as location-based services (e.g., ATM or store locations).

Accomplishing either of these goals effectively will require the expert assistance of those responsible for marketing the products, services, and ideas of your organization. This group of people will probably also include those in your organization most experienced in graphic design—a vital part of any Web application. They will be able to develop a visual look for applications that is consistent with the overall corporate image, and with other marketing media used by your organization.

Legal Advisors

Although it is not necessary to have a lawyer present throughout the planning process, some expert advice will be useful on several points. Most of these revolve around intellectual property issues.

- ❐ How can we widely distribute our information and still retain our copyright?
- ❐ If we produced this data under a contract, can we still distribute it to others?
- ❐ Are we allowed by law to charge money for this data, and if so, how much? (This is especially important for government agencies.)

❑ If we are charging for our information, should we license it or sell it outright?

❑ Are we infringing on someone else's copyright if we distribute this particular data set?

❑ Does the distribution of this data set infringe on anyone's personal privacy as protected by law?

❑ What are the ramifications of distributing or selling this data internationally (due to the global nature of the Internet)?

If you are in the business of distributing geographic or other information, you are probably very familiar with these questions and their answers. However, many readers of this book may be considering DGI as an appendage to their organization's mission, or perhaps as a new mission, and need to look into these issues.

Who Is Our Audience?

As with most forms of communication, it is vital that you know to whom you are distributing your information, so that you can tailor your services to their needs. The following are questions about your audience that require answers if you are to best satisfy the user of your information.

❑ *How much do they know about the subject matter your organization deals with?* If your organization works with wetlands, does your audience include scientists, activists, and politicians who also look at wetlands? Are you expecting the general public to visit, which on average knows little about wetlands?

❑ *How much does your audience know about cartography and/or GIS analysis?* Are they familiar with concepts such as buffering, overlay, projections, and map symbology, or will your application need to hide these operations?

❑ *Does your audience include those who have their own GIS software?* Will they be able to take your data and use it in their system, or will you need to provide all of the necessary functionality in your site?

❑ *What are the capabilities of the computers used by your audience: their hardware, software, and Internet connection?* Do they have the powerful computers and direct connection necessary to do graphics-intensive applications, or are they coming from home via a modem, where they will be leery about any graphics?

❑ *Is your audience within your organization, or out on the Internet?* Are they people you can train, trust, and give mandates to, or could it be just about anyone in the world?

❑ *How much access should your audience have to your information?* Do you have private or secure information to which you do not want any outside access? Who should be able to view or edit which data sets?

There are four main categories into which you should be able to classify most of your potential users (some people may fall into more than one category): internal personnel, GIS users, experts in your area of expertise, and the mass market. You may be interested in segregating, tailoring, and distributing various pieces of your information to one or more of them.

Internal Personnel

As discussed in Chapter 1, the intranet has become a very popular form of connecting individuals, departments, and organizations, in which standard Internet tools (such as the Web) are used not for the general public but for the personnel within an organization.

The principles of DGI are as applicable to an intranet as to a presence on the Internet.

This is because geographic information can be valuable to many departments and individuals in your organization, even though many, if not most, will not have the resources (e.g., time or computer power) or expertise to install and use a copy of your GIS software. The solution to this problem is a well-designed intranet DGI application, which will allow anyone in your organization (but nobody outside the organization) to have access to your GIS.

Personnel access your GIS with general purpose software, such as Web browsers, which they probably already have on their computers. As with any Web site, you can control exactly who has what form of access to which data sources. One drawback to this approach, however, is that it is very difficult (but possible with some software) to design a system that allows users to make changes to the database. Thus, this approach is most useful for infrequent users, or those who only need to look at the information, not alter it.

Internal personnel tend to know a lot about the geographic information of your organization (at least as it relates to individual jobs), but little about the workings of GIS. Therefore, applications for this audience could, but do not need to, explain the information in detail. However, the applications should have interfaces that are very easy to use, with documentation explaining GIS operations.

GIS Users

This second group are those people with access to the Internet who have their own GIS software and the expertise to use it. They may not know much about the work your organization does, but you might have information of value to them.

For example, if you are a civil engineering firm, you may produce digital orthophotos, terrain models, or street maps for the area around a site on which you are working. Once you are finished, other people living or working in the neighborhood (e.g., developers) may be interested in obtaining (perhaps purchasing) this information from you.

Because this audience already has the software necessary to use your information, an application geared toward them needs to distribute GIS data itself, rather than finished map products. They do not need GIS analysis or mapping capabilities, except perhaps as a reference to enter spatial queries for data sets and files that match the information they are looking for.

Experts in Your Area of Expertise

This audience includes those people and organizations on the Internet who are very knowledgeable about the area of expertise of your organization, but may not know much about GIS or have access to normal GIS software (of course, another category might be those with both subject and GIS expertise). For example, you might be an environmental advisory organization that does scientific studies (using GIS among other techniques) on environmental issues. Other environmental scientists may be keenly interested in your results, in the form of maps and reports, but may not have access to the same GIS software you used.

This category would also include the residents of a municipality, the major clients of a business, and organizations similar to yours with which you collaborate or even compete. For this audience, you do not need a host of auxiliary documents that give easy-to-understand descriptions of the subject, nor does the interface need to be simplistic. However, your application will need to provide all GIS and mapping capability necessary to communicate your results. Extra GIS capabilities may also be valuable here to allow other experts to analyze your data themselves or to explore the results in more detail.

Mass Market

The last major group you may be interested in sharing information with is "everyone else." Because of the global nature of the Internet, there will likely be thousands, if not millions, of people who are at least casually interested in your organization. They may not have much knowledge, or even care, about the internal workings of your organization, and will almost certainly not have their own GIS. However, they may be a very useful, and profitable, market for your geographic information.

Take, for instance, the previous example of an environmental organization. In addition to collaborating with the scientific community, this organization may also have a mission of educating the general public on its activities and its stand on various issues. There are many types of geographic-environmental information the public might be very interested in viewing, from the location of proposed wilderness areas, to the habitat ranges of endangered species, to the location of local toxic chemical sites. This environmental organization would be interested in making the connection between this information and its audience.

Because the mass market audience has little knowledge of either GIS or your subject area, it is highly unlikely they will be doing any powerful analysis of your information. You can therefore usually communicate your message with simple, predesigned maps and explanatory text.

Integration with Other Systems

This system will not operate in a vacuum. It is only one part of your general Internet site, which will likely include documents and tools unrelated to mapping. It is also a part of your GIS implementation (if you have one), which includes geographic data and the software for using it. The following illustration shows the relationship of an Internet site to a Web site and an overall GIS.

DGI services are part of both your Web site and your GIS.

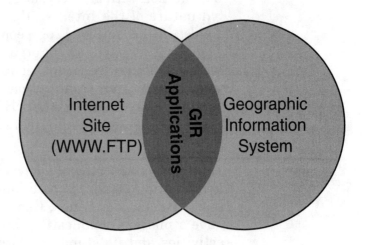

It is important that your strategy for a DGI application be an integral part of plans for your Web site and your GIS. If these systems already exist, your choices for implementing DGI will likely be constrained by the characteristics of these systems. If they are not in place, you should plan all three systems together to maximize consistency and compatibility. The following are some issues you should consider when planning your DGI applications in concert with your Web site and GIS, whether you have these in place or are planning to develop them simultaneously.

❒ *Software Compatibility.* The software you use for your DGI system, whether bought off the shelf or developed in-house, will act as an interface between your GIS software and your Internet server software. Thus, the software needs to communicate with both. Web server interface standards such as CGI, Java, ISAPI, and NSAPI (see the glossary) are common, and DGI software from any vendor will likely work with almost any Web server. Communication with your GIS is more difficult, and your options will usually be limited. Both of these linkages are discussed in greater detail in Chapter 5.

❒ *Interface Consistency.* The users who will be accessing your DGI applications will also be using the rest of your Web site. Thus, to aid the usability of the DGI service and improve its aesthetic quality, its design should be consistent with the design of the overall Web site. Use similarly styled elements—fonts, colors, and icons—in the maps and HTML pages that constitute the DGI interface.

❒ *Application Consistency.* It is helpful to make your DGI applications match the applications in the standard GIS. For example, if a municipal GIS has an application for selecting a parcel and obtaining a title history, a Web-based interface should contain the same function, and it should have a matching appearance. If your DGI applications are similar to your GIS applications, they will likely share at least some program code, thus saving you programming time. This is also important if the application you are designing will be used in an intranet. The consistent appearance and functionality will make training and maintenance easier.

What Should We Share?

One of the biggest decisions you will face when planning a DGI application is what information to make available, and in what form it will be best distributed. You will not want or need to make all of your organization's geographic information available to the public, in the same way that you do not want every bit of internal financial information up on your Web site. You need to assess the nature of and select from the geographic information you have. This selection process involves several issues, which follow.

The Public Nature of Your Information

Does the general public have a legal right to see your information, or is it up to your discretion what you give them? When you consider the accessibility of each of your data sets, there are three main categories under which information might fall: public domain, distributed with copyright, and private.

Public Domain

If you are a governmental agency in the United States, at least some, and possibly all, of your information is in the public domain, under the Freedom of Information Act for federal agencies and under similar state laws for state and municipal agencies, as described under "Because You Have To" in Chapter 1.

Under these laws, the information created by a government must be made available to the public at no more than the cost of distribution, except for that which infringes on personal privacy or national security. If you use the Internet, the cost per download is minimal, so most agencies distribute data for free.

Even if you are not mandated to give out information for free, you may choose to do so (that is, put your information into the public domain), depending on the goals of your organization and the type of information. A popular approach is to post quickly produced, simplified information as a "teaser," with higher-quality maps and data for sale if the user is interested (discussed later in this chapter under "Can We Make Money?").

Distributed with Copyright

This category includes information you clearly own, and distribute with certain limits on the rights of the public to reuse and copy it. Although the application of copyright law to the Internet is generally not well understood by users, it is generally accepted that downloading something is not a copyright infringement (even though you are making a copy). However, the copyright holder has rights over everything else (including personal use, modification, and redistribution), which the holder can relinquish, license, or retain. This is explained in more detail in Chapter 8.

You can freely distribute your GIS information, with use restrictions, or you may want to sell it (or license it for a fee) with few restrictions. For example, the static continental maps used in the Virtual Tourist (*http://www.vtourist.com/webmap/maps.htm*) are copyrighted. Free rights are granted for copying the maps and placing them in other noncommercial Web sites, but a small licensing fee is charged to commercial sites that include the maps.

Private

In most nongovernmental organizations, much of your information may not be for public viewing by any means. This does not mean that this information should be immediately eliminated from your DGI strategy, however. It may still be a valuable part of your intranet because other personnel within your organization may be valid users of various pieces of this information.

Using the foregoing categories, you should be able to look at the geographic information you have and determine which portions you want to make freely available to the public, which portions you wish to make available for a price, which you will share within your organization, and which will not be shared at all.

For example, a civil engineering firm may have converted local government GIS data for use in its own software, which it then chooses to make freely available. This is a perfectly legitimate practice because the original data resided in the public domain. Other products, such as city-wide orthophotos and high-precision terrain models, may be offered for sale. However, detailed site plans are not likely to be openly available on the Internet, although it would be valuable for engineers to be able to share them over a local network.

Forms of Distribution

The next issue to consider is how you will distribute the information you have selected. What type and what level of access does your audience need and deserve? In what way should you present your data sets that will be most useful to your audience? There are four basic forms or ways of making geographic information available in your DGI site: raw data, maps, searches, and analyses.

Raw Data

If your audience has GIS software capability, it is not likely they will want finished maps. They will want the same data you have, which they can put into their own system to analyze. Although the internal data structure of a GIS is generally too complex (i.e., many files and directories) to transport easily, there are several standard formats to which you can convert your data for transport (for which your users will need a similar conversion routine in their own software).

Common formats include Arc Export (.E00) by ESRI, DLG and DEM of the USGS, and the U.S. Spatial Data Transfer Standard (SDTS). Several other formats are commonly used for transporting the geometric portion (e.g., a line that represents a road) of GIS data, although attributes, such as the name or surface material of that road, are lost. The most common of these formats is the DXF format by Autodesk.

Because a pretty interface is not necessary (although it may be useful for helping users find the data they need), FTP has been a popular alternative to the Web for transporting this type of information.

One of the first examples of this approach has been the USGS's GeoData service, discussed in Chapter 1 (the data service can be found at *http://edcwww.cr.usgs.gov/doc/edchome/ndcdb/ndcdb.html*). This site allows users to find medium- and small-scale digital data, which can be downloaded using FTP. The following illustration shows a directory of USGS digital data sets.

*Directory containing
USGS digital data
available for download.*

Index of /pub/data/DEM/250/C/

Name	Last modified	Size	Description
Parent Directory	06-May-94 13:15	-	
caliente-e.gz	19-Feb-94 11:45	1M	
caliente-w.gz	19-Feb-94 11:46	1M	
campbellton-w.gz	25-Feb-94 12:57	1M	
canton-e.gz	23-Feb-94 03:38	1M	
canton-w.gz	23-Feb-94 03:45	1011K	
canyon_city-e.gz	14-Feb-94 16:11	2M	
canyon_city-w.gz	14-Feb-94 16:12	1M	
cape_blanco-e.gz	24-Feb-94 08:57	999K	
cape_disappointment-e>	25-Feb-94 12:57	79K	
cape_fear-w.gz	18-Feb-94 13:11	52K	
cape_flattery-e.gz	25-Feb-94 13:56	1M	
carlsbad-e.gz	18-Feb-94 11:40	1M	
carlsbad-w.gz	18-Feb-94 11:51	1M	
casper-e.gz	24-Feb-94 08:45	1M	
casper-w.gz	24-Feb-94 08:45	1M	
cedar_city-e.gz	19-Feb-94 11:44	1M	
cedar_city-w.gz	19-Feb-94 11:45	1M	
centerville-e.gz	23-Feb-94 07:59	923K	
centerville-w.gz	23-Feb-94 08:04	1001K	
challis-e.gz	14-Feb-94 16:12	2M	

Maps

By far the most common way in which geographic
information is distributed is through maps. The map is
a good medium for communicating to both the expert
and the novice, and is therefore found in DGI applications geared toward all of the audiences previously
mentioned.

There are a variety of approaches you can take toward
incorporating maps in a DGI site. The maps may be predesigned, containing a symbology (e.g., colors, patterns,
fonts) you have determined beforehand to achieve a
very specific design. On the other hand, you may allow
the user to have some control over the appearance of
the map, by choosing which types of features will be displayed, or even changing the symbology.

The maps may be static, which you have drawn to certain specifications and that are not alterable on the site (e.g., a world map of population density). Alternatively, you might choose to make maps dynamic, where the map itself or the viewing perspective of it can be changed by the user, such as with a zoom-in/out feature for seeing a region of particular interest.

Format is another issue to be dealt with. On the Web, graphics are found primarily in raster (i.e., pixel or grid-based) formats, such as GIF and JPEG. Although the raster format works for many mapping applications, in some cases a vector (i.e., object-based) format may be more optimal. To date, no vector graphics formats are supported directly by any common Web browsers, but plug-ins (additional software attached to the browser, which is explained in more detail in Chapter 5) are available for some formats, including DXF, CGM, PostScript, and ShockWave (which supports animation as well as static graphics).

The difficulty with using plug-ins is that not everyone in your audience will want or be able to install them. Another issue related to map distribution is that you need to be able to generate maps from your GIS and get them into the desired format. Some software does this as a part of its programming, but others require an intermediate conversion program.

A good example of a DGI site that focuses on map-based applications is the street map browser, exemplified by the TIGER Mapping Service (*tiger.census.gov*) discussed in the first chapter, MapQuest (*www.mapquest.com*) from GeoSystems Global, and Vicinity's MapBlast (*www.mapblast.com*). These services contain a GIS of streets, points of interest, and various other features for the entire United States (and to varying degrees of detail for parts of the world), which users can browse using zoom and pan tools to get detailed maps of any location in the United States. The following illustration shows a MapQuest application.

The MapQuest Interactive Atlas, a map-based application by GeoSystems Global.

Simple Searches

Many of your users will be content to just look at the information you present; others will want to ask some questions. The most commonly desired query capability, and easiest to implement, is a search for a specific item in your database based on a set of criteria. These criteria can be either spatial (at a given location) or thematic (a feature with given attributes), or both.

Spatial criteria are usually entered either through a clickable map or in a form using geographic coordinates (e.g., latitude/longitude) or place keywords (e.g., state, county, city names). These forms can be used for entering thematic criteria as well. (These interface tools are explained in Chapter 6.)

The search is then performed using normal database techniques, rather than GIS software. The records that match the criteria are then returned to the user, either in map or text report format.

At least two examples of this application are common today. One is called a data clearinghouse, many of which are being created by many U.S. government agencies to aid in distributing GIS data sets as part of the NSDI. The user enters spatial (e.g., "Covers Los Angeles, CA") and thematic (e.g., "Annual traffic flows") criteria for the data they need, and the service returns a list of data sets that meet that criteria. The user can then view metadata (a detailed description of the data, including its scope, accuracy, currency, and so on, described in more detail in Chapter 6) for each data set to determine its appropriateness. This may even include a map showing the area the data covers. If the data is what the user wants, the data set can be downloaded directly.

Another example is the online telephone book, such as BigBook (*www.bigbook.com*)—an entry from which is shown in map form in the following illustration. Users are able to search a nationwide yellow pages for a specific business by location (city), name, and category. Matching entries are then listed, each of which can be shown on a map. Using specialized network GIS software, even driving directions from your location can be given. Even though the original search has little to do with GIS or mapping, it is still a spatial search application and is thus part of DGI.

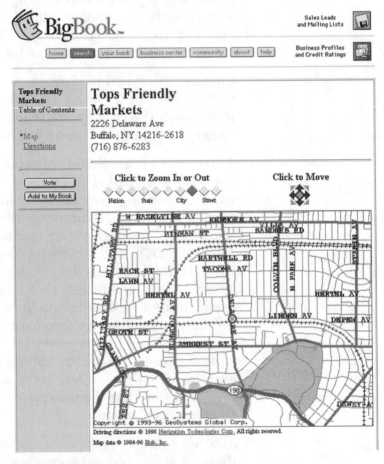

A map of a yellow pages entry in BigBook.

GIS Analysis

Depending on your organization and audience, you may wish to give users access to the full analysis capabilities of your GIS (or some subset thereof). This would allow them to perform complex multi-theme queries, create buffers and customized maps, perform statistical spatial analysis, and other tasks. Only rarely would you want people to be able to modify your data (i.e., in an intranet application). This type of service allows the user to create new data sets from their own analyses without altering the data you maintain. The data sets the user created could be stored on your server for their future use (perhaps with a time limit, after which the data sets are deleted), or you might offer the option of downloading their results as a map, report, or raw data set.

If you choose to provide this form of service, be prepared for some programming. Today's Web-GIS software does not put your Internet users directly into the standard interface of your GIS software (command line or GUI). You will need to design a custom Web interface to your GIS applications; you may even need to modify or recreate those applications to work with the Web server.

GRASSLinks, discussed in Chapter 1, is a well-known service providing this level of information. Another good example is the City of Oakland (California) Map Room (*http://ceda.ci.oakland.ca.us/index1.htm*). This site uses ESRI's MapObjects Internet Map Server (discussed in Chapter 4) to create an interactive map browser for the city's GIS, including parcels, streets, and aerial photographs. When viewing parcels, the user can query a particular plot and get complete attribute information, as shown in the following illustration.

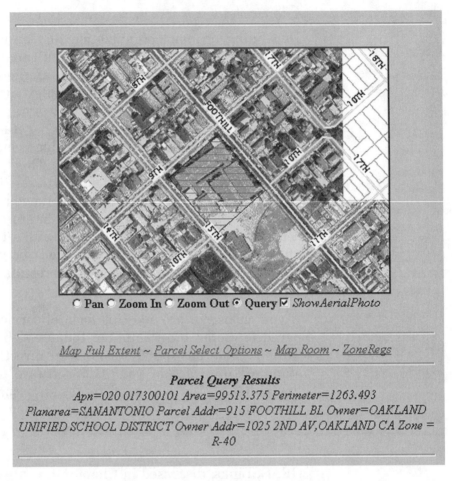

A map from the City of Oakland, with attributes from a selected parcel (hatched at center).

Any GIS application could include one or more of these forms of access, depending on what your major goals are and what level of access you wish to give your audience. After you have figured out what you have to offer and whether or not you have the means to make it available, you need to assess its potential profitability, the subject of the next section.

Can We Make Money?

This question is certainly important for businesses interested in a profit from their data, but it should also be considered by government and nonprofit organizations that have the right to charge at least a small amount for information. The following are reasons you may wish to charge people to obtain your data. Each may lead to a different pricing strategy.

❑ *Profit-making Venture.* Many companies are in the business of producing geographic data for the purpose of selling it. This includes cartographic data such as streets, demographic and marketing statistics, and custom-built GIS analysis products. The Internet can now be used as another marketing tool and retail outlet for selling your data to a global market.

❑ *Resale of Existing Data.* Your organization may produce geographic data as part of projects for clients who pay for the production costs. It is very likely that other people or organizations may be interested in obtaining this same data, and even pay for it if necessary. If the terms of your contract allow you to retain copyright on this information, you may be able to do whatever you like with it, including selling it to other parties. This can easily be a money-making venture because you have already covered the cost of producing the information. Any sales represent profit.

❑ *Cost Recovery.* Whether or not your data was produced for profit, you may have incurred very large costs in producing it, and distributing it may even be a liability. A Web server, software, and Internet connection all cost money. However, by selling the data for an appropriately low price, you may eventually be able to recoup much of these costs. In most organizations, you can set a price that will, after a certain expected number of sales, pay for the cost of producing and distributing the information. Any additional sales represent either profit or funds available to reinvestment in the production of information. As previously mentioned, if you are a government agency in the United States, you may only recover the cost of distribution, but this provision might include costs associated with implementing and maintaining the server and connection.

As you think about these three motivations, you may decide to charge for at least part of your geographic information. Which data and at what price is entirely up to you, but there are several factors to consider. These include the expectations of your market and its willingness to pay, the pricing structures of any possible competitors, the actual costs involved, and the time over which you recover those costs.

Approaches to Online Sales

So how exactly will you charge people for downloading and using your geographic information? The main issue to consider is the sensitivity of the Internet market. Because users can quickly move from one site to another, there is an expectation of ease and immediacy. Any unexpected obstacles (however small) will dissuade a large portion of your market. You need to use a method that will allow users to quickly find and purchase what they want, without a lot of hassle. There are three general approaches to setting up a purchase mechanism: products purchased directly, site subscriptions, and indirect sales.

Direct Purchase of Products

With direct purchase, users simultaneously pay for and receive the products of your GIS (i.e., maps, raw data sets, reports). These may be prebuilt items you put into an online catalog, or items that can be assembled dynamically according to the purchaser's specifications. The latter might include general reference maps or the results of an online GIS analysis.

For this strategy to be successful, you will need to carefully consider a good price for these products. This is more difficult for dynamically generated items than for the stock items for which you already have prices. One difficulty to offering online shopping is the low individual cost of many of these items (e.g., a tabular report from one step of a GIS analysis, for which you may be able to charge only a few cents). Paying for items by credit card is still the most popular method of shopping online, and users tend to limit purchases to larger, higher priced items, such as entire data sets.

There is some inefficiency to this approach because the buyer is usually required to type in billing and delivery information in an order form area every time an order is placed. This favors the casual user, who comes to the site only once (or very rarely) and buys a few things at once. However, as electronic cash schemes become more prevalent, both of these problems will be solved for users who transfer small amounts of money because electronic cash allows for the transfer of money with almost no overhead.

Subscription to Site

For frequent users, who will be using your site regularly, a subscription approach may be better. Here, each user has a unique "account" on your system, which they will need to enter (by username and password) every time they access your site. All of the billing and delivery information is entered when the account is created. When a user subsequently logs in, this information is retrieved and associated with new purchases logged.

Members are then billed regularly (usually monthly) for their usage over the time period, either by a flat monthly rate or based on an accounting of exactly what they used or purchased the previous month. The latter works better for larger-priced items (e.g., entire data sets), whereas the former works better for a large number of smaller-priced items (e.g., query results), for which a detailed accounting would be too much of a problem. They might also be billed at the time of purchase, as with the previous approach, but without the hassle of entering their billing information every time.

This is essentially a bulk-discount concept, similar to the difference between using a pay phone and having your own line. In the previous approach, each order has a set nuisance overhead (e.g., entering billing information). In this approach, that overhead is concentrated into the initial subscription activity, and the day-to-day use of the site incurs no overhead. However, the subscription process is more involved than a single purchase (e.g., waiting for your account to be created); therefore, infrequent users will not subscribe. A compromise solution may be to use a direct purchase approach, with a subscription system as an option for your most frequent users.

Indirect Sales

In this approach, the GIS is not the object of the purchase but is used as a shopping tool. The products and services you sell with this approach could be anything with a locational aspect, such as real estate, fast food delivery, and restaurants. Here a map might be a useful tool for users to select the exact location at which they wish to access the product or service.

In fact, the catalog items being sold might even be maps or GIS data, separate from the free maps used for the selection process. This is the approach employed in ImageNet (*www.coresw.com*) by Core Software Technologies, who sell satellite imagery using a map browser for selecting a region of interest and for displaying the spatial footprint of data sets for sale. The following illustration shows a display of an ImageNet query.

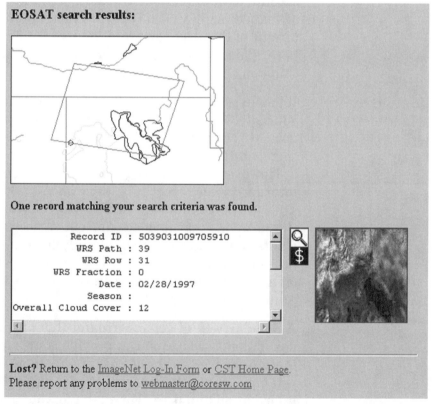

Results of a query on ImageNet, with a map showing the coverage of the image.

An example of an indirect sales approach might be a company that sells stock landscape photographs. Buyers might be interested in a certain style of landscape (e.g., a tall building), but they are more typically interested in pictures of a particular area (e.g., the Grand Canyon). A map would be invaluable for showing the locations of photographs you have in stock, as well as for users to specify, by clicking onscreen, the area in which they are interested as part of their search.

Each of these sales models has been used profitably as part of both general and DGI Web sites, but you must take the Internet culture (your intended market) into account to be successful. In addition to the expectation of immediacy previously mentioned, another characteristic of this user community is even more important here. This is the general misconception that everything on the Internet should be free.

If you expect users to pay for all of the information on your site, you will have very few customers. They will not pay for anything without some idea of its value to them, and they can only estimate that value by seeing it. However, the technology of the Internet is such that in order to see something, they have to download it. This is the conundrum of Internet marketing. Once the user has downloaded an item to see if he or she wants to purchase it, they already have it, and so do not need to pay for it. How do you prevent being ripped off as seller? How do you at the same time attract and protect sales?

A popular way of getting around this apparent paradox is the "teaser." You can provide some simple information for free to attract users and show them what you have, at which point you offer them higher quality information at a price. For example, you might be selling customized street maps. A free map browser lets users find the location they want, showing them very simplified, low-resolution raster images. For many users this will be good enough, and thus you will reach a wide market. If someone wants a higher quality map to print out, or wants you to calculate the fastest route between two locations, or wants your raw GIS data for a given area, they can choose to purchase it at any point.

This approach is used in the ImageNet site previously discussed. A generalized version of each satellite image is displayed, which is too crude to actually be useful, but good enough for customers to know whether it is exactly what they want before they purchase the full image.

Who Should Manage the Site?

The last major issue to deal with in the early planning process is the manner in which you will set up and manage your DGI applications. This issue will be most important when it comes time to allocating staff and capital resources to the application (as described in Chapter 5). Basically, you are determining who will do what work. You may want to contract out some phases of the implementation or some parts of the application, with the rest of the work done in-house. There are three basic approaches to this allocation process: do everything yourself, hire DGI consultants and Web designers, or use a Web hosting service.

Do Everything Yourself

In this strategy, you are responsible for all of the resources necessary for the design, implementation, and maintenance of your DGI applications. Each of these tasks will require skilled staff, powerful computers, and a direct connection to the Internet.

This approach is best applied when you have staff with expertise in this area, or if you plan to do several, or very specialized, applications. If your organization already has an established GIS and an existing Web

site, you probably already have the necessary personnel and computer resources. If not, you may need to
hire new people, or reassign skilled people, as well as
purchase hardware. Chapter 5 focuses on planning the
resources needed for this option.

Hire DGI Consultants and Web Designers

Since few of us have the artistic talent necessary to
produce an appealing yet professional Web site, Web
design services, which employ graphic artists (at least
the good services do) who build Web sites, have become
very popular. You contract with these services to create HTML pages and graphics, which you then store
on your own server. Generally, you maintain the site
from then on, although you may also contract with
them to keep it up to date. Either way, you are still
responsible for maintaining the server hardware and
software, and thus need to devote the time of capable
staff to your site.

In the same vein, there are a few consultants who are
experts in integrating GIS and mapping into the Internet, with whom you can contract to help develop (and
perhaps maintain) your DGI applications. Although
many of the GIS software vendors (such as Intergraph
and ESRI) have their own consulting services, there
are also many third-party companies that can create
an entire system for you and deliver it ready to go. If
you do not want to go to that extreme, many of them
can also work with you in more of an advisory role,
helping you create your own site.

Use a Web Hosting Service

You may not have the resources to do any of the project
in-house. High-bandwidth Internet connections and
capable staff can be very expensive, especially if you

are only planning a small site. Many organizations are in this situation when it comes to Web sites in general; therefore, Web hosting services have become popular.

These companies employ a large Web server to simultaneously run sites for several organizations. Although the computer and the data are not at the client's physical location, users cannot tell the difference. Because the host is responsible for the day-to-day maintenance of the computer, all the client has to do is update the information, which can mean considerable savings for a small site. These services are also usually coupled with a Web design service, saving the client from having to develop the visual appearance of the site as well.

However, there is a problem when it comes to DGI applications. These almost always require specialized server software, as well as a GIS, and very few (if any) hosting services have, or are even willing to install, this software. They rarely have expertise in this field, and therefore are not able to maintain the software or the information very well. Thus, for GIS-related applications, you may have no choice but to keep the server at your location and maintain it yourself.

Summary: The General Plan

The various issues and approaches described in this chapter should help you and your planning team develop a general plan of the DGI applications you want to develop. However, you will almost certainly require more detailed information to actually get the project approved and started. You will need to put together a budget proposal, identify exactly which personnel and computer resources will be devoted to the project, and create a detailed design of the applications, including specifications for the necessary software.

Information to help you develop the details of, and implement, your plan is found in the remainder of this book. The next chapter takes a more technical look at strategies for developing DGI applications. Subsequent chapters look at specific software you can use, as well as implementation and maintenance issues you can anticipate as you develop your site.

3

A Host of Solutions

Available Designs for DGI Sites

Now that you have an overall strategy for the types of geographic information you wish to incorporate into your Web site, it is time to get down to details. For any strategy you have, there are a variety of programs and approaches available from which you can choose. This chapter outlines the specific designs that can be used in your DGI applications.

Server Versus Browser Processing

Your DGI application will likely require a great deal of computer processing. Requests must be processed, searches performed, GIS analyses carried out, reports generated, and maps drawn again and again. Where should this work be carried out? The client/server model upon which the Web is based allows for the sharing of processing between client and server, but does not dictate the exact ratio.

There is software available for both Web browsers and servers that can handle at least some of the computing involved in DGI services. You will need to decide which elements of this you will perform on your own machine and how much you will defer to your users' computers. There are many factors that could influence your decision either way. These involve the capabilities of your users and their computers, your market, and the Internet connection between them.

Heavy Server/Light Server

The fundamental motivation behind the client/server model has been to concentrate certain data and software on a single machine, from which clients can access them. A single centralized data/software site is generally cheaper, easier to keep updated, and uses computer power more efficiently than distributing everything. Access to information is also easier to control and facilitate.

The same is true of the Web. By keeping GIS data on a central machine, you can update it more easily than if users have their own copies scattered all over the world. Because they are always using your one set of data, they will always have the most recent version. Keeping your clients' GIS functionality up to date should follow the same principle. In the world of the Internet, the appearance and functionality of sites change frequently. Forcing your users to download a new version of your complete GIS software every two or three months would drive many of them away.

Concentrating the processing workload on a primary server (referred to as a *heavy* server) has other advantages. By using a very powerful server, not as much load is put on your users' computers, which likely have a wide variety of capabilities. Nothing is more annoying than visiting a Web site that instantly uses up your CPU time and memory.

This concentration of work has its downside, however. GIS and mapping are processor-intensive tasks— much more so than the simple disk-to-network file transfers performed on most Web sites. GIS software tends to require a powerful machine even for normal work. If you create a popular site that has ten to twenty requests processing simultaneously, the load can be overwhelming in terms of processing time or even the ability to process at all.

Many server-heavy sites, such as the Xerox Map Server and the Census Bureau's TMS, both discussed in Chapter 1, have had serious problems keeping up with their own popularity. When the processor gets overloaded, requests can take considerably longer to process (three times the normal response time or longer), or the server may stop responding to new queries. TMS now runs on four server computers concurrently (delegating requests between them automatically, as discussed in Chapter 5) to achieve a reasonable response time: about three to four seconds to draw each map.

Another drawback is that if all processing is done by your server, you will need to spend time and money developing a GIS-capable server. This involves interfacing GIS software and data with a Web server to process live, dynamic requests. It also usually involves building custom applications that allow users access. Even with the commercial interface software now available, this remains a daunting task.

The last problem with server-heavy DGI applications is that Internet communication and server processing is required for even the most menial of tasks. Every time the user zooms in on a map, a new request is made to the server, and a new map is generated from scratch and returned to the user interface. This increases network traffic and response time significantly. It also makes your site less responsive and interactive, less useful, and ultimately less popular.

Thick Client/Thin Client

Normally, Web browsers fall into the category of *thin clients*; that is, most processing is handled by the server, and all the browser has to process is the display. For most browsing tasks, this is entirely adequate. However, because a thin client requires a heavy server, DGI and other client server applications may not always work well under this model, as described in the previous paragraph.

The alternative is the *thick client*, a setup in which the client does the majority of the processing. If better graphics handling and even GIS capabilities were part of the browser, it could process many tasks, such as map browsing (e.g., panning and zooming) and spatial query formation (e.g., outlining an area of interest on a map) without bogging down the network or taxing the server.

For obvious reasons, browser developers are not willing to add these capabilities to their software for such a small market as the GIS community. To remedy situations such as this (everyone has their own special type of data to display), browsers have begun to include support for Java, ActiveX, and "plug-ins," which add functionality to the browser on demand. When users come to the site, they can download small programs called *applets*, which then become part of the browser software. This allows GIS developers to create their own spatial processing applets.

A Java or ActiveX applet can be made to download automatically when people visit your site, but the download will not be persistent; that is, they must download the applet again the next time they access the site. On the other hand, a plug-in must be installed and downloaded like a normal program before your site can be used. However, once it is installed in a user's browser, it is available permanently.

Several GIS and mapping sites have taken the "plug-in" approach. Intergraph's ActiveCGM plug-in is a relatively simple viewer for vector files, but it provides some important capabilities, such as panning, zooming, and linking map objects to URLs (hence the term *Active*CGM).

In addition to these capabilities, vector graphics formats such as CGM tend to represent vector GIS data much more efficiently (in fewer bytes) than raster formats, such as GIF and JPEG, which are supported natively in most Web browsers. This means that files can be downloaded more quickly; thus, your site will be more responsive. Vector formats also allow for the manipulation of individual elements, such as hyperlinking and selection.

Other sites provide more functional applets, which further alleviate the server from work. The Java and ActiveX versions of MapQuest (*www.mapquest.com*) are basically more responsive versions of the HTML-only interface. The MapGuide plug-in from Autodesk (see the following illustration) provides a wide variety of map viewing functionality, such as panning, zooming, object attribute reports, and turning the display of layers on and off. More advanced applets, which include basic GIS query functionality, are also available.

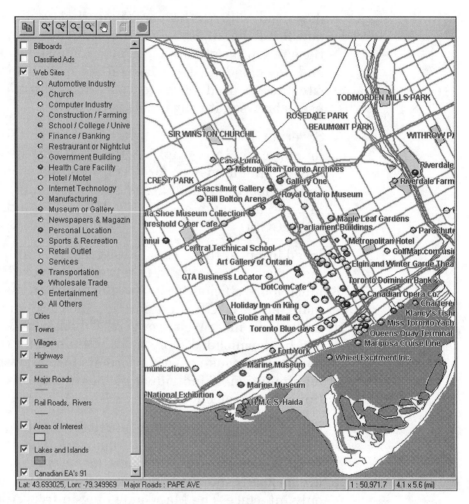

The MapGuide plug-in, which includes many map browsing capabilities.

Giving the client more responsibility for processing data and queries almost always reduces the amount of data that must be transferred over the Internet, because more efficient raw data can be sent rather than graphics files. Also, the client can do many operations itself, rather than having to send a new request for every operation, with the server returning a new image.

The main drawback to this approach is primarily in the downloading and installation of the applet. Depending on its capabilities, the file size of the program may reach a megabyte or more. No matter how automatic you make the process, a large proportion of your potential audience will not be willing to spend the time to download this file. This is largely because the online community expects convenience. If something cannot be viewed immediately, many people will move on to the next site.

This inconvenience factor is even greater for a plug-in, which requires a user to exit the browser, install the plug-in, and restart the browser. Thus, by using an applet or a plug-in, you will likely limit your audience to those serious users who are either very interested in your spatial information or use it frequently.

Finding the Client/Server Balance

Do you build your DGI service with very little processing performed by a light server—most of it being delegated to a thick client—or do you use a standard Web browser as a generic thin client, with the majority of the work done by a heavy server? The answer depends on the many factors discussed in this section. Essentially, you must find your own balance in the trade-off between breadth and depth.

A thick client (interacting with a light server) allows for flexible and powerful analysis, but cannot handle as much network traffic, limits your audience, and can be a nightmare to maintain (i.e., keeping all of the clients up to date). A thin client/heavy server setup uses a considerable amount of bandwidth sending so many maps, and is generally limited to more simple applications. However, it can be used by many more people, including those not willing to download and install a thick client just to use your service.

In general, a thin client is best for mass market applications, with a potential audience of thousands or millions of users that have little demand for advanced GIS analysis capabilities. Thick clients work well for services used by a smaller set of frequently visiting GIS-savvy users.

Four types of client/server balance are summarized in the following table. Columns represent the possible solutions to the separation of client and server tasks, whereas the rows represent the tasks accomplished by the client and server in each solution. These tasks include displaying the map images on the client's screen, simple browsing tasks such as pan and zoom, formulating and/or executing simple spatial queries ("What is here?") by pointing to a location on a map, drawing maps from raw GIS data, and employing full GIS analysis capabilities such as overlays and buffers.

	Thin Client	**Balanced**	**Thick Client**	**GIS Client**
Server Tasks	Map browsing Query Analysis Map drawing	Query Analysis Map drawing	Analysis Map drawing	File serving
Transfer	Raster maps	Raster/vector	Vector maps	Raw data
Client Tasks	Display	Display Map browsing Query input	Display Map browsing Query	Display Map browsing Query Map drawing Analysis

Examples of the thin client approach previously discussed include the TIGER Map Server (*http://tiger.census.gov*) and GRASSLinks (*http://regis.berkeley.edu/grasslinks*). The Java and ActiveX versions of

MapQuest (*http://www.mapquest.com*) are a good example of a medium-weight client and server. Thick clients include the MapGuide browser (*http://www.mapguide.com*). The last category consists of full-featured GIS programs that can use data directly from the Internet. (This is explained in more detail later in this chapter under the heading "Net-savvy GIS Software.")

Types of DGI Applications

The types of sites on the Internet incorporating GIS and mapping are distinguished by characteristics other than the client/server distinction. In fact, there are at least seven types of DGI services from which you can choose. The ideal solution to your site configuration strategy depends on several issues, including those raised in the previous chapter, such as the form of your information and the analysis capabilities you wish to provide. The GIS skills of your intended audience and the software available to them will also play a crucial role.

Most of the following seven general applications have been used in many places, but a few have only been proposed. The commercial software packages discussed in the next chapter vary as to their applicability, but most of them are appropriate for a few of these applications. The following applications are presented from least to most complex and difficult to create and use.

Raw Data Download

In the first application, there is actually very little processing being done by either side. The server is basically delivering files, the architecture of which is shown in the following illustration. The data sets come directly from the GIS, although they are usually packaged (e.g., ARC/INFO Export files) and/or reformatted (e.g., into standard transfer formats such as SDTS or DIGEST).

Architecture of a raw data download site.

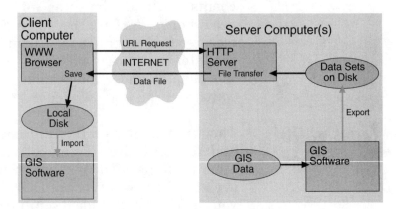

The data sets are then placed in a network-visible file system, usually on either an FTP or Web (HTTP) server. The folder or directory structure is usually organized to aid navigation, such as by state or theme. The user may be able to browse the directory structure and download files of interest, or may be granted access only to particular files (such as in a data sales operation, after the user has made a purchase).

The basic assumption of this approach is that your users will have their own GIS software with which they can use your data. This software has nothing to do with the download process; the GIS work is done "off-line," after the data sets have been put on the user's local disk. (The gray line in the previous, and subsequent, illustrations represents this offline access.) This will obviously limit your potential audience to GIS-capable users, but for many organizations that produce data, this is exactly the audience they are trying to reach.

One of the best examples of this technique is the USGS node of the NSDI (*http://nsdi.usgs.gov/nsdi*), especially the GeoData Online section discussed in Chapter 1. In addition to simple text navigation tools, the site also provides map and search tools for finding data sets appropriate to the user's needs. Map and search tools are actually separate DGI techniques, which are described in the sections that follow.

Static Map Display

This approach is probably the simplest in terms of the work done by both the server and the client. The geographic service consists basically of predesigned map images (either raster or vector). The map graphics are created using the GIS or graphics software normally used to create maps, but saved in a raster or vector format that can be viewed on Internet browsers. This usually means GIF or JPEG raster formats, and CGM, DXF, or Shockwave vector formats, although the latter set requires users to install plug-in viewers.

Once an image file is created, it is distributed on the Web like any other graphic image, and viewed in the user's browser either as part of an HTML document (using the <IMG> tag) or by itself. Thus, no geographic processing takes place at either the Web server or the browser; to both sides, the file is just another GIF file (or some other format). Although it is technically simplistic, this approach will likely have the widest audience because it requires the least amount of browser and user capability. The architecture of this type of display system is shown in the following illustration.

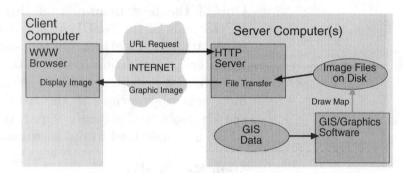

Architecture of a static map display system.

A popular service using this method is the Perry-Castañeda Map Library at the University of Texas (UT)-Austin (*http://www.lib.utexas.edu/Libs/PCL/Map_collection/Map_collection.html*). As a federal repository site, this library has a wide variety of public domain, government-produced maps, including a set of simple maps of countries from the Central Intelligence Agency.

UT has scanned a large number of maps as JPEG images, which are freely viewable from the directory in the Web site. This type of site is very useful and popular for the majority of the general public, who rarely wish to perform spatial analyses or other tasks that would require a full GIS capability.

Another site, with a little more functionality, is the Virtual Tourist (*http://www.vtourist.com/webmap*). The maps here are also static graphic images, but using *imagemaps* (a standard Web technique documented in most HTML guides and HTTP server manuals), they have been made interactive. Thus, when the user clicks on the map of Europe, the pixel coordinates of the cursor are spatially compared to a database of hot spots, each of which contains the outlines of a country. In this case, the outlines are entered manually; no live GIS processing is involved. (Techniques for using image maps with a GIS are discussed in Chapter 6.)

Along with each outline is a URL for the respective country. Therefore, when the polygon is found that encompasses the selected point, the user is automatically forwarded to a site for that country.

Even though this site is interactive, it belongs in the static map category because nothing spatial is being performed beyond the default capabilities of HTTP servers and browsers. The drawing of maps and tracing of area outlines for the link database are done as separate processes because the spatial comparison is performed by the HTTP server, with no unique GIS capabilities. As opposed to some of the GIS engines described in the material that follows, here, the map drawing and the creation of the hot spots are distinct functions.

This same idea (and often the exact same set of maps) has been duplicated in many sites. Maps are used to show the global offices of multinational corporations (e.g., Solid Systems CAD Services, *http://www.sscs.com*, as shown in the following illustration), the location of the nearest store (e.g., Lechters Housewares, *http://www.lechters.com*), tourist destinations (e.g., Resorts Online, *http://www.resortsonline.com*), and in almost any other application where location is important but is not the focus of the organization. The map-based interface to each type of data in the USGS GeoData site described in Chapter 1 (the navigation interface only, not the destination data sets) is another good example of interactive static maps.

Solid Systems CAD Services, a non-GIS Web site that uses static maps to aid navigation.

Metadata Search

The metadata search technique is roughly analogous to a library catalog in which the holdings are GIS data sets rather than books. It is basically a database lookup application, but with the ability to do simple spatial queries. The database consists of *metadata*, which are structured descriptions of geographic information available from an organization.

Each available data set is described by a *metadata record*, which contains a predefined set of fields describing various characteristics of the data set that may be of interest to users. Common metadata fields include those that detail the subject matter (e.g., vegetation, roads, and pipelines), projection, coordinate system, physical file format, source and accuracy of information, and spatial footprint (i.e., the geographic area covered by the data). (Metadata is discussed in more detail in Chapter 6.)

The data sets themselves are not in the database, but may be available via one of the other types of services listed in this section, with the metadata including the URL address at which the data set can be obtained. However, online availability is not necessary, and the metadata may alternatively include a telephone number or address users can contact to obtain the data set. It might also include a URL pointing to an online order form.

The most prominent example of this type of site is the National Geospatial Data Clearinghouse (NGDC; *http://fgdlerhs.er.usgs.gov*), one of the primary parts of the NSDI. This service, the architecture of which is shown in the following illustration, acts as a central search interface to any number of metadata databases, which are scattered around the world. Most of the current databases index raw spatial data from U.S. federal and state government agencies, which can be downloaded as described previously in the "Raw Data Download" section. However, other private and international organizations have registered servers that deliver information using any of the types of services described in this chapter.

Architecture of the National Geospatial Data Clearinghouse, a metadata search service.

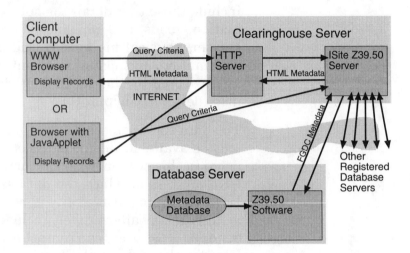

To use the service, the user enters desired criteria (which will almost always include a spatial region of interest) into a query form. This is sent to the clearinghouse server, which forwards the query to all registered database servers using Z39.50, a standard Internet protocol for performing database searches.

Each server processes the query against its own database, returning the matching metadata records to the clearinghouse. This machine immediately collates and formats the results, returning a list to the user. Each metadata record may contain only text, or it may contain text as well as a link to such things as a secure order form, other Web pages with more information, or the data set file itself.

The Clearinghouse also has a more advanced Java-based interface, which includes an interactive map browser (a service in the Dynamic Map Browser category), as shown in the following illustration. This map interface is invaluable for allowing the user to enter spatial criteria in an intuitive manner (e.g., drawing a box on the map) and for displaying the spatial footprint of each of the resulting data sets (from the information in the metadata record, wherein the spatial footprint is stored).

The Java search form for the National Geospatial Data Clearinghouse.

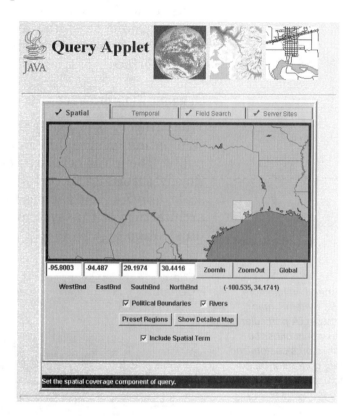

Similar metadata search services with both form and map-based interfaces include Core Software's ImageNet (*http://www.coresw.com*), which is based on the same centralized query/decentralized database concept as the NGDC, as well as the Alexandria Digital Library Catalog (*http://alexandria.sdc.ucsb.edu*) from the University of California at Santa Barbara.

Related to this type of service are those that perform simple spatial and thematic queries on other types of non-GIS databases. An example of this type of query was given previously about a stock photography company with a landscape archive. Information about its available photographs (location in latitude and longitude, as well as place name, date and time, format, subject) could be stored as metadata in a database, which could be queried using text fields as well as a map interface.

Dynamic Map Browser

This category is probably the most popular way of disseminating geographic information to the general public. The maps here are not static but are drawn on the fly according to certain parameters, such as the scale, location, and themes to be included. The general architecture of this system is shown in the following illustration.

General architecture for a dynamic map browser. A particular site would either be based on a customized map generator (upper right) or a DGI gateway program (lower left).

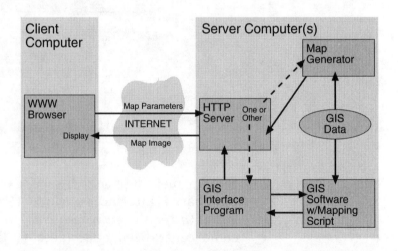

The maps may either be delivered by a standard GIS (e.g., MapInfo) through a DGI gateway program, or generated by a map generator, which is a program written specifically for this purpose. Either way, maps are drawn from one or more spatial data sets. Most current sites, such as EtakGuide (*http://*

www.etakguide.com, shown in the following illustration) from Etak, are based on the thin client approach previously discussed.

The server draws the map (using preset colors and other symbology) as an image, and the browser displays the image. Whenever the user wishes to change the view (by panning, zooming in and out, placing labels, turning themes on and off, and so on), a new request is sent to the server with slightly different parameters than the last map. The server immediately generates a new map with the new parameters, which is returned to the user, restarting the cycle. (The technical details of this interface are discussed in Chapter 6.)

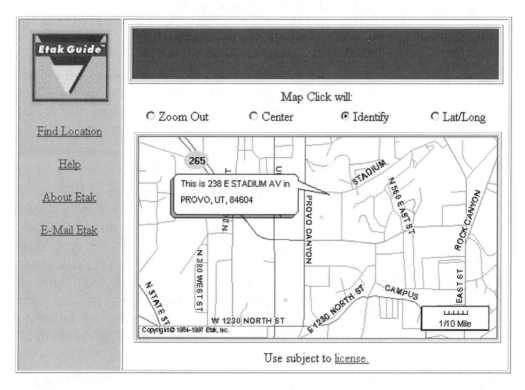

Etak's EtakGuide, a dynamic map browser.

As discussed in the previous section, this approach can reach any Internet-capable audience. It is also adequately responsive and interactive. However, it uses a great deal of bandwidth and puts a very heavy load on the server, which may have to generate several maps simultaneously for different clients.

The alternative is to use a medium-weight map browser client, such as Autodesk's MapGuide. Here, the maps are actually drawn by an augmented client (in the Autodesk example, a Netscape plug-in), which has some control over such visual elements as which layers will be drawn.

The server is responsible for delivering semiprepared data on demand, including coordinates representing desired themes for the area within the view window, as well as in the vicinity of the view window so that the user can zoom out. The server retains control over most of a map's appearance, including colors, line styles, and labeling. Potentially, thicker clients could be created that would give users even more local control over map appearance, including colors, line styles, and labeling.

Data Preprocessor

Although very few examples of this type of DGI application currently exist, it is potentially a very useful service for users with their own GIS software. Many low-end GIS programs, such as AtlasGIS and Map-Info, do not have extensive capabilities for preprocessing spatial data, such as converting from various formats or changing projection. A data preprocessor, the architecture of which is shown in the following illustration, can assist with some of these functions.

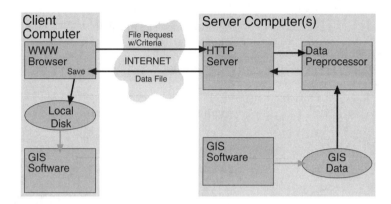

Architecture of a data preprocessor service.

The purpose of this application is to augment a raw data download service, described at the beginning of this section. Rather than simply delivering data as is, some form of rudimentary processing may be done on the data before it is delivered. The most useful applications here include reformatting a data set into the native format of the client's GIS software, and changing the projection and/or coordinate system to match that of the user's own GIS data.

Although it sounds useful, thus far it has been implemented very few times in real Web sites. One program that can be used to build this type of site is the Feature Manipulation Engine from Safe Software (*http://www.safe.com*). This software, discussed in the next chapter, has an associated demonstration (*http://www.safe.com/fme/livedemo.html*) that is a good example of this type of service, shown in the following illustration. The native data is in MicroStation DGN format, but the site lets you select a file format useful to you, as well as the themes in which you are interested, and it converts the data on the fly into that format as it delivers it to you.

A demonstration of Feature Manipulation Engine.

▲ **Forestry Design File -> GIF/Shape/SAIF Live Demonstation**

After you fill out and submit the below form, the FME will translate the chosen portion of the input design file into the chosen format.

Available Output Formats

● GIF ○ SAIF ○ MapInfo MIF/MID NEW ○ Shape

Available Linear Feature Types

Miscellaneous	Roads	Water Courses
☐ Powerlines	☑ Highways	☑ Creeks
☐ Pipelines	☑ Secondary Roads	☑ Rivers
☐ Railways	☑ Logging Roads	☐ Snags
☐ Abandoned Railways	☐ Trails	
☐ Flumes and Ditches		
☐ Glaciers		
☑ Seismic/Cut Lines		
☑ Dykes		
☐ Ferry Routes		

Available Polygonal Themes

○ Ownership   ○ Planning Cell   ● Forest Region
○ Biogeoclimatic   ○ Operability   ○ P.S.Y.U./T.F.L.
○ Timber Supply Area ○ ALR   ○ Provincial Forest
○ Range Units   ○ Inventory Region   ○ *No polygonal theme*

[Translate] [Clear Form]

Web-based GIS Query and Analysis

Probably the most advanced applications available today give users a wide range of functionality normally found in GIS software. Using either traditional Web tools such as HTML forms and GIF image maps, or newer techniques such as Java, the user has an interface with which to perform various queries and analyses against your GIS. This architecture is shown in the following illustration.

Architecture of a Web-based GIS query and analysis service.

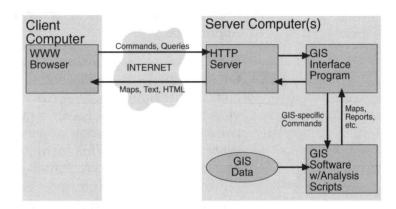

The functions you can let your audience use include almost anything a GIS can do, such as attribute queries, spatial queries, buffers, overlays, classification, map display, and even data editing. You have complete control over which of these operations will be available, as well as which data sets will be visible. You may wish internal users to have almost complete control, but your Internet audience to have access only to rudimentary visualization tools.

This type of service almost always requires quite a bit of programming, because the user does not have direct access to your GIS (in very few cases would you want them to). This means you will need to build a customized interface to the specific operations to which you want users to have access. This will involve both the creation of Web interface tools that allow for the formation of queries and analysis requests (e.g., forms, and maps you can point to) and the creation of scripts in your GIS that can process those requests and output the results. (the interface options are described in Chapter 6, and Chapter 7 includes information on the scripts.) Some commercial DGI programs have example applications, but you will need to do at least some manual configuration to adapt any strategy to your application.

Good examples of this application include the GRASSLinks service discussed in Chapter 1, and For-Net (*http://www.gis.umn.edu/fornet*) from the University of Minnesota. Fornet's map interface, shown in the following illustration, allows for public browsing of state forestry data. This includes themes such as vegetation types, land use, transportation, and even satellite imagery; the interface allows users to display any or all of these layers on a map.

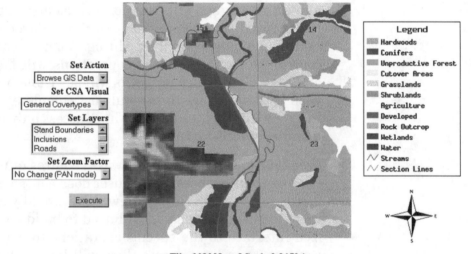

Tile t103109 and Scale 1:26734

[Reset] [Help] [Comments]

The map interface of ForNet.

Although its analysis tools are not as extensive as GRASSLinks, ForNet does have a good query feature, where users can obtain extensive attribute information for any displayed object. It also includes a tool for viewing satellite imagery, including the ability to compare vegetation index images from multiple time periods.

This site is based on the thin client model, although a Java-based medium-weight client is under development at the University of Minnesota. In fact, most of the current services in this category are server heavy because of the difficulty of programming powerful GIS capabilities in a customized client.

Net-savvy GIS Software

The last type of application uses an extremely thick client. Basically, the client software is a standard GIS or desktop mapping program that has the added ability to use live data from the network. The server in this case does nothing but deliver data in real time. The architecture of a net-savvy GIS is shown in the following illustration.

Architecture of a net-savvy GIS based on NFS.

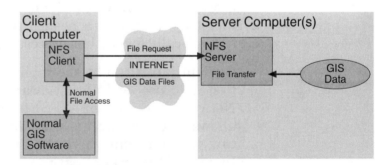

Net-savvy GIS software is one of the primary goals of the Open GIS Consortium (OGC; *http://www.opengis.org*), which is a collective of GIS vendors, developers, and other parties whose mission is to increase the ability of disparate GIS platforms to operate together. The consortium's vision of GIS in a distributed computing environment (DCE) includes GIS clients, scattered across an intranet or the Internet, that can communicate with one another and use shared data. These Internet-savvy GIS clients must have at least three primary capabilities.

Remote File Access

The first category discussed in this section (raw data download) requires users to first download a data set to a local disk before the data can be used. In contrast, this type of software would be able to read remote data sets as easily as it reads local ones.

This is very efficient, both in terms of the local disk space conserved, as well as bandwidth, because only the data needed for a particular task is transferred, rather than the entire original data set. Also, data sets can remain in their native format, without needing to be packaged for easier download. The drawback is that transferring information over the Internet is considerably slower than reading it from a local disk. This technique, therefore, would only be useful in certain cases, perhaps in which the user accesses the remote files only to extract a needed portion, which is then saved locally for further processing.

Currently, this can be accomplished in two ways. The first is to use an Internet-aware file system such as Network File System (NFS) from Sun. This service allows you to connect to disk drives (with authorization) residing anywhere on the Internet and treat them as if they were local.

In contrast, most other network file sharing techniques, such as NetWare and Windows Networking, cannot see outside your local network, although Microsoft is releasing a somewhat Internet-aware version of Windows Networking called distributed File System (dFS). NFS client and server software is available for almost any platform (it is usually built into UNIX, but must be purchased for other operating systems). As shown in the previous illustration, once an NFS drive is mounted, any GIS software can access the remote data without even knowing it is remote.

One popular site that uses NFS to distribute live data is the CD-ROM Information System (*http://ucdata.ber-keley.edu/cdrom .infosys.htm*) at the University of California at Berkeley (formerly housed at Lawrence Berkeley Labs). This service contains over 250 CD-ROM disks, including all of the data available from the U.S. Census Bureau, as well as several other federal public domain data sets, including NIMA's Digital Chart of the World. All CD-ROM disks are mounted on a group of NFS servers, which can be freely accessed by anyone on the Internet.

In addition to these network protocols, which use a generic file system metaphor, technologies are being developed under the direction of the OGC that are focused specifically on the transfer of geographic information across the network. Geographic data objects (GDO) is an extension of the OLE and COM standards from Microsoft to better handle spatial data.

The Open Geospatial Datastore Interface (OGDI; *http://www.las.com/ogdi*) is based more closely on standard Internet protocols. Both of these potential standards are based on a client/server architecture called the *Open GIS Model* (OGM), which is shown in the following illustration. On the computer with the data is a data server (not a standard ATTP server), which is specifically designed to read a certain data format. The Web-savvy GIS has its own client module that can request data from the server.

Understanding Various Formats

Because data may be coming from a wide variety of computers and software, a net-savvy GIS program needs to be able to read data that may be stored in many different forms other than the native data format of the software. This aim is beginning to be realized in newer GIS software such as ArcView from ESRI, which can natively use data files from many sources. The range is still limited, but they are heading in the right direction.

This is also part of the mission of the OGC (*http://ogis.org*), an industry organization that aims to increase the ability of disparate GIS platforms to work to-gether. The OGM standards previously mentioned include transport mechanisms and a common file format used in transfer. Software based on these standards, such as Intergraph's GeoMedia (based on GDO) and LAS's GRASSLand (which uses OGDI), support data in any format for which there is a corresponding data server.

Real-time Projection and Positional Matching

Real-time projection and positional matching allows you to compare and combine data sets from different sources—data sets that may be in very different projections and coordinate systems. Whereas most GIS programs have projection operations, these generally convert one file to create another in the desired projection. A net-savvy GIS would be able to do it on the fly, projecting points as they are drawn, while leaving the files intact.

Currently, very few GIS programs can accomplish this, including Intergraph's GeoMedia (see the following illustration). When themes are loaded from various sources, and a desired projection is selected for the display, the projection of each data set is determined (if possible) and the elements are converted as they are drawn.

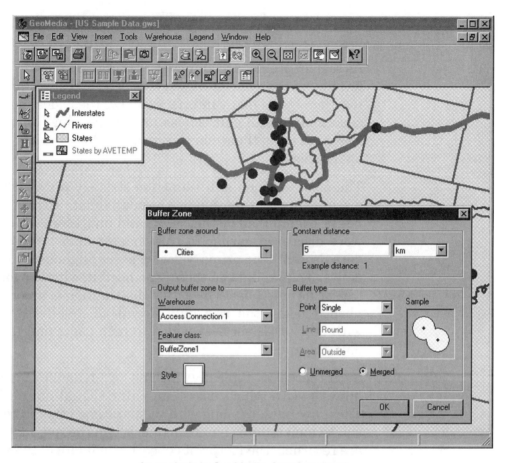

An analysis in GeoMedia, from Intergraph.

Obviously, the market for this type of service is much more limited than any of the other categories previously discussed, because not only must users have their own GIS software, the software must be Internet savvy. Thus, the first major audience for this will be GIS professionals, who frequently need public data from remote sources, data from several fee-based services, or data on an intranet with the data stored on many servers.

Combination Services:
Clearinghouses and Digital Libraries

As seen in previous examples, it is common for a site to have more than one of these types of services working together. Using many interfaces allows the widest possible audience to gain access to your information, as well as making that information useful for a variety of applications.

The most common multi-function strategy previously discussed is represented by the spatial data clearinghouse, exemplified by the NGDC and ImageNet, which were discussed earlier in this chapter under "Metadata Search." The primary goal of a clearinghouse is to deliver spatial data to those who need it, whether for free or for purchase. The actual distribution is accomplished using the raw data download architecture, with other DGI services added to help users find the data they need.

Such services typically include a dynamic map browser with which users can enter spatial queries. These are passed (along with thematic and other criteria) to a metadata search engine. When the metadata are returned, the footprints of each candidate data set can also be displayed using a dynamic map browser. From the metadata record, users can follow a link to download or order the actual data. Thus, three separate DGI services—data download, dynamic map browser, and metadata search—are combined to create a useful service.

Another combination service is the digital library, such as the experimental Alexandria Digital Library (*http://alexandria.sdc.ucsb.edu*) being developed at the University of California at Santa Barbara. This concept tries to embody many of the functions of a real library.

The clearinghouse "card catalog" is a major part of this digital library (and the only part implemented at Alexandria so far), but the complete library includes other functions. Library users will eventually be able to peruse and research the materials within the facility if they do not have their own software; this necessitates dynamic map browser and Web-based GIS applications. In fact, many developers envision intelligent "librarian" agents that will automatically gather, analyze, and display spatial information to answer complex requests such as, "Tell me about the historical growth of Santa Barbara, California."

Summary: Choosing a Strategy

In the first three chapters, a wide range of strategies and methods were suggested for delivering geographic information on the Internet or an intranet. This variety is an asset, not a disadvantage, because it allows you to tailor your site to accomplish exactly what you wish.

As you plan and implement your DGI strategy, you will need to continually consider the issues presented so far. In fact, you may offer several related services, each of which requires its own strategy and set of answers to these issues. The questions that will help you choose a strategy might be summarized as follows:

❏ Why do you want to distribute your geographic information?

❏ Who is the intended audience or market for this service, and what are its characteristics and expectations?

❏ What parts of your GIS database do you want to share with this audience?

❏ Can you, or do you wish to, charge for the use of your data?

❐ What types of display, query, and analysis capabilities do you want your audience to have?

❐ How should the interface to your information look, and what user capabilities should it offer?

❐ Should processing take place mostly at the server, mostly at the client, or somewhere in between?

❐ Will your site be built, housed, and maintained within your organization, or will you delegate some of these responsibilities to specialist companies?

As you deal with these questions and issues, you should be able to develop a clear, implementable design for a service that will be highly useful to your desired audience. The following are a few hypothetical applications that can be derived from answers to the foregoing questions.

❐ A consulting company has data sets they wish to sell to other GIS users.

Strategy: A clearinghouse service, with a metadata search capability and a map browser for input of spatial criteria and display of spatial footprints. Data download is not available, but metadata points to a secure order form.

❐ A city wants to increase public accessibility to municipal spatial data.

Strategy: A dynamic map browser, using themes stored in the city GIS (e.g., streets, water bodies, parks, emergency facilities, and parcels). Simple GIS query tools would be included, such as looking up the owner of a parcel.

❏ A nonprofit organization wishes to use detailed maps as part of political literature, showing where the habitat of an endangered species is being encroached.

Strategy: One or more predrawn static map images of the areas of interest. Raw data sets available for download by very interested parties.

❏ A group of academic geographers scattered around the world wishes to collaborate on a research project requiring extensive GIS analysis.

Strategy: Each researcher would have an Internet-savvy GIS, which would access a shared data server or even communicate with all other researchers in real time. (A Web-based GIS analysis service might work, but might not provide the flexibility and functionality required.)

During the development phase, it is a good idea to peruse existing sites to get ideas for creating your own, as well as to be aware of potential competition. Many sites have been listed in the preceding chapters, but a more complete guide can be found at the companion Web site for this book at *http://www .geog.byu.edu/gisonline.*

Implementing this plan will rarely be easy. Thus, the next chapter looks at DGI tools, developed by commercial and nonprofit organizations, that you may be able to use to create your services.

4

Commercial DGI Software

A Wide Variety of Solutions

In the early days of the Web, building online GIS and mapping sites was a daunting task. Many of the early services, such as the Xerox Map Server (*http://mapweb.parc.xerox.com/map*) and the TIGER Mapping Service (TMS; *http://tiger.census.gov*), were custom-built generating programs using proprietary data formats. Others, such as the Virginia County Interactive Mapper (*http://ptolemy.gis.virginia.edu/gicdoc/mapper/tigermap.html*), interfaced with a standard GIS platform. However, even in the latter case, the gateway had to be custom built, and because the GIS software of that time was not made to handle such real-time connections, the services were slow and not very useful.

This requirement for extensive programming limited the DGI realm to just a few sites, most of them either research prototypes or well-funded commercial products. For a wider variety of services to be created, they would first need to be much easier to build.

In answer to this need, most of the major GIS software vendors, and many other companies, have introduced DGI software. These programs generally provide high-level tools for creating online mapping and analysis services, as well as the engines for running these services. The intent of this chapter is to discuss these offerings, to help you decide which, if any, you should use for your projects.

Difficulties in Comparing Software

Many details—such as price, speed, and feature lists—about the products described in this chapter are not included. Neither will you find head-to-head comparisons of products. This is because packages vary widely in their approach, capabilities, requirements, and ease of use. Many are designed for very different applications, and use different data sources. Therefore, given a particular strategy you might employ, it is likely that only a few will apply, and comparing them directly is nearly impossible.

Another reason for the omission of certain details and comparisons is that most of these programs were created relatively recently. Therefore, their features, performance, and pricing are likely to change more frequently than the edition cycle of this book.

The intent here is to introduce each package, and discuss its general usefulness for the particular strategies outlined in the previous chapters. The companion Web site to this book contains more up-to-date information about each piece of software, as well as more detailed comparisons of their functionality.

It is anticipated that you will find a few programs particularly appropriate to your situation, and evaluate their detailed features according to your needs. For any strategy you might adopt from Chapter 3, one or more of the programs discussed in this chapter should suffice. The exceptions are the strategies of raw data download and static map display, which do not require special, spatially enabled software.

Although available packages vary widely, they can be grouped into seven broad categories according to their general approach and architecture. Each category exhibits strengths and weaknesses for each of the service types discussed in Chapter 3. Some categories will work best for one strategy, some for another.

The groups discussed in the material that follows are focused enough so that their constituent programs can be evaluated in terms of a particular service. This chapter is therefore organized into these seven categories: remote map generators, proprietary-data map generators, GIS-data map generators, live GIS interfaces, Net-savvy GIS software, non-Internet solutions, and spatial catalog servers.

Remote Map Generators

The options in this category may be the best for those who have no GIS software or data, and who want only to include a few basic (but dynamic) maps in their Web sites (see the discussion of the "map browser service" in Chapter 3). Each of these services is a map browsing site in its own right, generating maps from its own data according to parameters given in a URL. These parameters include the themes to be shown, the real-world area to be included, the size of the output image, and sometimes even a color scheme.

If you know the syntax of the parameters, you need not use their browsing interface because you can use a customized URL to directly request the map image you want. This image-generating URL can be inserted into an HTML page in the same manner as a URL for a static image. For example:

```
<H2>Here is my house</h2>
<img src="http://tiger.census.gov/cgi-bin/mapper/map.gif?lat=37.09&
lon=-113.56&wid=0.120&ht=0.043&iht=350&iwd=400&mlat=37.07&
mlon=-113.56&msym=blupin&mlabel=My+House"><p>
```

When the Web browser retrieves the HTML page and displays it, it makes a separate HTTP request for the image, as it does for normal images. However, instead of returning to your server to download a static file, it sends the request to the map server, which generates the image and returns it to the browser for display in the document. The HTML would generate a document similar to that shown in the first of the following illustrations. The subsequent illustration shows the architecture of a remote map generator site.

HTML document with a map generated using a TIGER Mapping Service map.

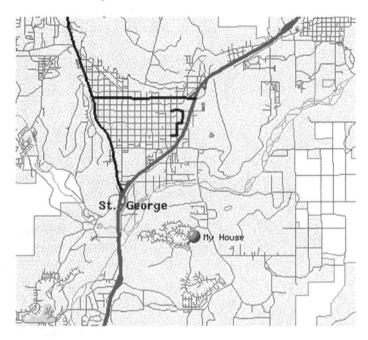

Architecture of a remote map generator site.

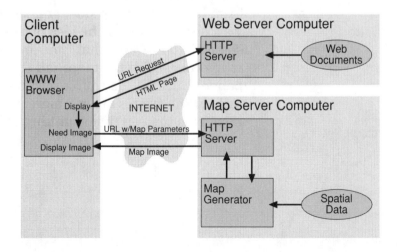

With this approach, you can even create your own map browser interface, including adding buttons for panning and zooming. When someone hits one of these buttons, an image is requested again, but with slightly different parameters for generating the map.

The greatest advantage of this approach is that you, as the site builder, do not need to download, install, or maintain any special software or data. To add your own spatial information (e.g., store locations), you can plot point locations on the maps in most of the remote map generator services, described in the material that follows.

For the most part, the maps you can generate are limited to the United States. Although available, data for the rest of the world is typically very coarse. In addition, analysis capabilities are limited to address location and possibly routing. Therefore, this category applies to organizations that want a few maps, but nothing complex.

Another potential downside to this approach is dependence on the remote mapping engine. If that engine, over which you have no control, has any problems (e.g., it becomes very slow, or goes down temporarily), your site will suffer. The idea of this category is to hide somewhat the fact that you are not generating the maps yourself. If the maps fail to work, you will likely be identified as the source of the failure. Fortunately, most of these sites are commercial, and have sufficient resources to keep them running efficiently.

EZ-Map! by Etak

```
http://www.etak.com/etakMoD
```

This service generates high-quality (relative to the other services discussed here) street maps of the United States rapidly. Use is restricted, however, because it is a commercial service; Etak requires that all uses, although free, be registered with them (*http:www.etak.com/etakMoD/registration.html*), and that you include a link back to their site. This site allows you to save maps to your own disk, making it easier to deliver static maps, such as the one shown in the following illustration.

A sample map from the Etak engine.

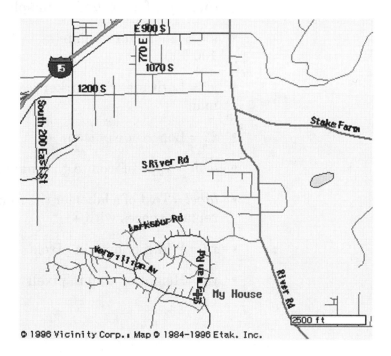

☐ **Data Source**. The service is based on Etak's well-known proprietary U.S. street database, which is essentially an improved version of the Census Bureau's TIGER data.

☐ **Available Themes**. The maps show streets and highways, streams and water bodies, railroads, parks, and points of interest.

☐ **Parameters**. The URL to generate a map image is:

```
http://www.mod.etak.com/gif?CT=y1:x1:s&IC=y2:x2:icon:label&W=x3&H=y3
```

- *y1* = Latitude of center of map in decimal degrees (positive in the northern hemisphere)

- *x1* = Longitude of center of map in decimal degrees (negative for western hemisphere)

- *s* = Scale factor of map (as in 1:100,000 when s = 100,000)

- *y2* = Latitude of a marker icon to be placed on the map

- *x2* = Longitude of icon

- *icon* = Type of icon (e.g., Smallcross)

- *label* = Text of a label for the icon, in URL form (i.e., replace spaces with +)

- *x3* = Width of map in pixels

- *y3* = Height of map in pixels

MapBlast! by Vicinity Corporation

```
http://www.mapblast.com
```

Vicinity allows you to include in your Web pages any maps generated by MapBlast, provided you furnish a link back to the service. Vicinity's maps are of a fairly high quality (almost identical to the maps from Etak), and have the unique feature of several color schemes from which to choose. Unlike EZ-Map, Vicinity does not give permission to save the maps for use in your site.

❐ **Data Source**. Uses Etak street data (see the previous discussion of EZ-Map).

❐ **Available Themes**. Same as for Etak's EZ-Map.

❐ **Parameters**. The URL controls the color scheme, map size, and coverage area, and allows for custom markers. The parameters for the map are not published, but are similar to those for EZ-Map. When you browse their service, and get the map you want, you can click on "Add map to your home page" to generate the appropriate HTML fragment, which can then be copied to your own page.

MapQuest LinkFree by Geosystems Global

```
http://www.mapquest.com
```

In addition to high-quality U.S. street maps, this site also has worldwide map data, although not of the same detail as the U.S. data (only major highways are shown). With this service, you do not include the map in your own Web page; you include a hypertext link that points to the MapQuest service and a particular map or maps, such as the one shown in the following illustration.

A sample map from the MapQuest engine.

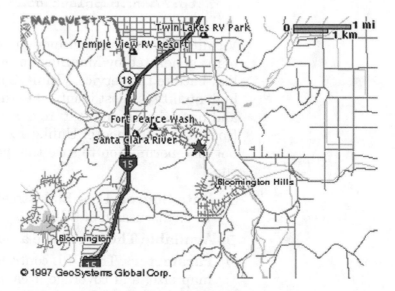

© 1997 GeoSystems Global Corp.

As with the Etak service, MapQuest requires that you register your use. MapQuest also offers MapFree, a service that allows you to save maps to your local computer for use in static pages.

☐ **Data Source**. Geosystems has its own proprietary street database of the United States (combining TIGER and other sources). The worldwide data comes from Geosystems' own database, Navtech, and other sources.

☐ **Available Themes**. United States: cities, highways, streets (with addresses), parks, federal lands, rivers, and points of interest. World: cities, highways, and rivers. You can add your own markers and labels to maps from both services.

❐ **Parameters**. You can control the map size and coverage area, and place points on a map. See the Web site for details of the parameters required to make a custom map.

MapQuest Connect/InterConnect by Geosystems Global

```
http://www.mapquest.com
```

This service essentially lets you build your own map browser. Alternatively, you can simply include maps in your Web site, the images of which are generated at the MapQuest site. Unlike some of the other services listed here, Connect and InterConnect are available only for a fee.

InterConnect has one extra feature: your site is advertised in your organization's location when anyone looks at your area in MapQuest. The MapQuest site lists several current clients of these two services, which you can review to see if they will work for you. One of these clients, BigBook, is shown in the following illustration.

BigBook is a popular service built using MapQuest Connect.

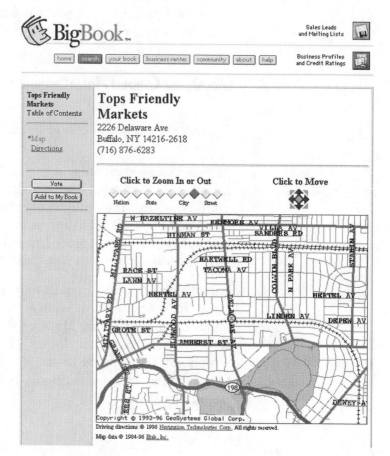

- ❏ **Data Source and Themes Available**. See MapQuest LinkFree.
- ❏ **Parameters**. When you purchase either service, MapQuest gives you a custom scripting toolkit, which you can incorporate into your own Web applications to include dynamic maps.

TIGER Mapping Service by the U.S. Census Bureau

```
http://tiger.census.gov
```

One of the first services in this category to be developed, TIGER Mapping Service is still popular. With four servers concurrently processing maps, the site tends to work with reasonable speed. It will create thematic maps from Census statistical data, a unique feature among the services in this category. Because it was developed as a noncommercial proof-of-concept, and because it uses raw TIGER data, the map quality is not at quite the same level as the other services discussed in this chapter.

A sample map from the TIGER Mapping Service.

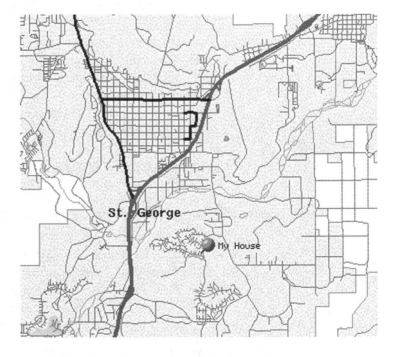

❑ **Data Source**. Most of the data comes from the 1995 version of TIGER, which has been reformatted into a binary format (not TIGER/Line) for real-time map drawing. An extract from the Summary Tape Files (statistical data from the 1990 Census) is used for thematic maps. It also allows you to submit one or more points to be displayed on the map.

❑ **Available Themes.** The server accesses all categories of data contained in TIGER: states, counties, cities, highways (labeled), streets (no labels or addresses), some parks, congressional districts, railroads, tracts and block groups, rivers, water bodies, military sites, and Indian reservations. The site also includes thematic maps of several popular statistics from the 1990 Census, such as income and race by tract or block group.

❑ **Parameters**. In the URL, you can control map size, coverage area, themes displayed, thematic map variables, and placement of your own points. A legend graphic can also be obtained with the same parameters. Details on parameters can be found at *http://tiger.census.gov/instruct.html*.

Vicinity Interactive Maps by Vicinity Corporation

`http://www.vicinity.com/vicinity/vim.html`

The Vicinity Interactive Maps (VIM) service provides all of the tools you need to build your own map browser service. The maps are generated by Vicinity's MapBlast engine, as previously described, but with a wide variety of customization options available.

Like MapQuest Connect, VIM is a commercial product that must be purchased. The price includes licensing and all tools. An example of this service is the Internet Yellow Pages site (*http://yp.ameritech.com*) offered by Ameritech. This site is shown in the following illustration.

Ameritech's Internet Yellow Pages site uses Vicinity's VIM service.

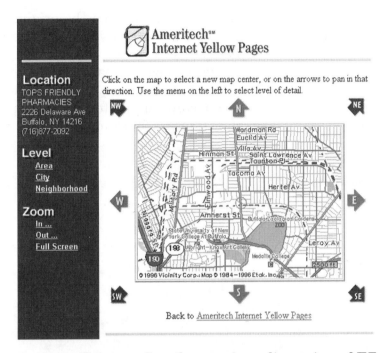

□ **Data Source**. See the previous discussion of EZ-Map and MapBlast.

□ **Available Themes**. See the previous discussion of EZ-Map and MapBlast. Also allows you to add your own list of locations to a map.

□ **Parameters**. Because it is a complete browsing service, you can control the positioning and size of the maps, as well as the accompanying browser tools.

Xerox Map Server by Xerox PARC

```
http://mapweb.parc.xerox.com/map
```

Maps generated from the original Web mapping site can be included in your Web site without obligation. The maps cover the entire world, but their quality is not very good compared to the other services in this category. It can also be rather slow because it is not extensively supported by Xerox. A sample map is shown in the following illustration.

A sample map from the Xerox Map Server.

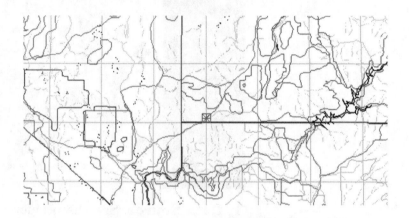

☐ **Data Source**. The world data has been converted into the engine's native format from the World Data Bank, a public domain file from the CIA. The higher-resolution U.S. data is converted from public domain USGS 1:2,000,000 digital line graphs.

☐ **Available Themes**. For the United States: coastlines, county and state boundaries, rivers, highways, railroads, and federal lands. For the world: coastlines, international borders, and rivers.

❏ **Parameters**. The URL can control the map size, coverage area, projection, themes displayed, and points (unlabeled) placed on the map. Details about the available parameters can be found at *http://mapweb.parc.xerox.com/mapdocs/mapviewerdetails.html*. The crucial option is to finish the URL with "/format=.gif," which will generate the image you can include in your HTML, rather than a link to their HTML page.

Proprietary-data Map Generators

These programs are essentially self-contained GIS software connected to the Web server. When parameters from a browser are passed to this type of generator, a self-contained database is used to generate a map or perform a simple analysis. The resulting map image and any text are passed back to the server, which forwards it to the browser. This architecture is shown in the following illustration.

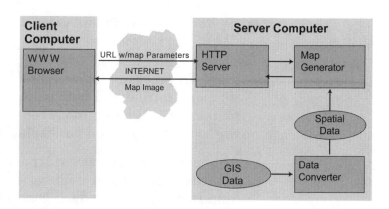

Architecture of a proprietary-data map generator.

The TIGER and Xerox services previously mentioned are examples of this type of engine. However, unlike the software discussed in this section, those services were not designed to be distributed to other parties.

The data used by proprietary-data engines is stored specifically for the DGI application, and is not used by any other programs. It is either prepackaged with the generator software (such as the national street map engines) or first converted from a native GIS format. Because the format is often designed specifically for this type of application, the data sets are typically very efficient in terms of processing speed and storage space.

Although the proprietary nature of this type of service can be cumbersome for organizations wishing to distribute their existing GIS data, it is very convenient for organizations that have no GIS but wish to include maps in their Web sites (such as the Visa ATM Locator found at *http://www.visa.com*). The most common use of this category is in dynamic map browsers.

Caris Internet Server by Universal Systems Limited

```
http://caris0.universal.ca
```

This program was the first commercial Internet mapping/GIS program; unfortunately, it was ahead of its time and intro-duced before anyone was looking. The product is still being improved and maintained, and works reasonably well for displaying GIS data.

Although there are several modules for performing database queries and the like, the core product is the Map Image module, which generates maps from GIS data. The most recent version of the software includes a Java interface (shown in the following illustration), which enables vector map browsing and query.

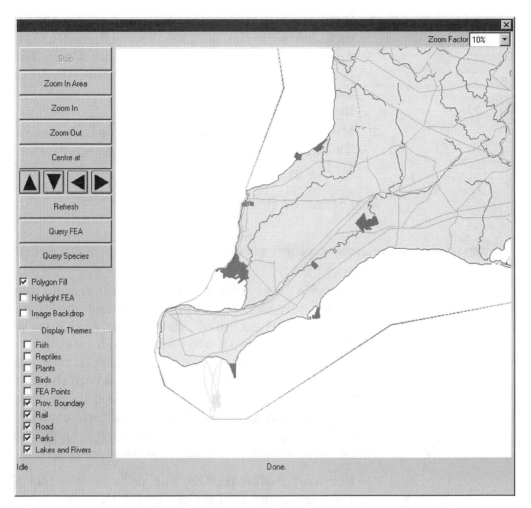

The Java interface used with Caris Internet Server.

❐ **Form of Software/OS**. It is a complete program for UNIX. The interface is command line driven, controlled by configuration files.

❐ **Web Server Interface**. Uses CGI.

❑ **Data Included/Formats Supported**. No data is included with the package. The software uses a proprietary format (the same as its Caris GIS product), but conversion tools for AutoCAD, DLG, and ARC/INFO are provided. Can also use raster image data.

❑ **Map Format**. Generates color raster GIF images. The Java version uses a proprietary vector format for transferring map data.

Map Server by Etak

http://www.etak.com/Products/webmap2.html

This service is very similar to those in the remote map generator category. The engine generates street and road maps of the United States from Etak's cartographic database, with a relatively high quality and rapid processing. The major difference is that the engine is at your own site. Etak sells you both the data and the map generator, which you install on your server. You also need to design your own interface around the maps. Additional modules are available for performing address matching and route finding.

❑ **Form of Software/OS**. The package consists of the engine and CGI scripts for UNIX.

❑ **Web Server Interface**. The engine is accessed by your Web server via CGI.

❑ **Data Included/Importing**. There are two versions of the service. The "basic" service includes TIGER street data, whereas the "premium" service includes Etak's augmented database. Both of these data sets are in a proprietary format, and converters for it are very rare; therefore, you will probably not be able to add your own GIS themes. You can, however, add point locations to the maps.

❏ **Map Format**. Maps are generated as color raster GIF images.

MapGuide by Autodesk

`http://www.mapguide.com`

This package consists of three programs. MapGuide Author, shown in the following illustration, is a tool to help you design the online maps to be generated from your data. MapGuide Server is the engine that stores and delivers map data. MapGuide Agent is an interface between the HTTP server and the MapGuide Server.

The server is multi-threaded (i.e., it can process several requests at once), and has security capabilities for restricting access to certain themes. To be scalable (i.e., to handle increasing load), maps can include themes from many agents. The software can also interface with database software (through ODBC) to obtain attribute data.

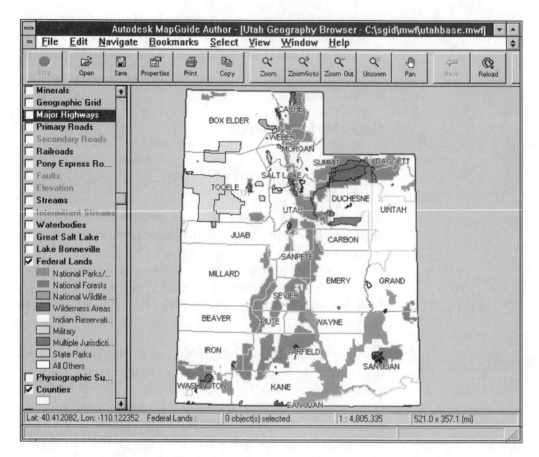

Designing an online map with MapGuide Author.

❐ **Form of Software/OS**. Both pieces are complete programs for Windows NT, with GUI interfaces.

❐ **Web Server Interface**. MapGuide Server can use all three major server interfaces: NSAPI, ISAPI, and CGI.

❐ **Data Included/Importing**. No data is included with the base product, although you can design maps that include your own data. The vector data format used by MapGuide (called SDF) is proprietary, but MapGuide Author can import from several programs, including AtlasGIS, AutoCAD, ArcView, and MapInfo.

❐ **Map Format**. Maps are transferred in MapGuide's proprietary vector format, readable only by the MapGuide plug-in. The plug-in automatically handles panning and zooming, including loading scale-dependent layers when necessary.

GIS-data Map Generators

As with the previous category, these engines are self-contained programs for making maps. In fact, they work in almost the identical manner, except for one critical factor. The programs in this category are able to read native GIS data. That is, when generating a map or GIS query, they look directly at files in the GIS. The architecture of this type of map generator is shown in the following illustration.

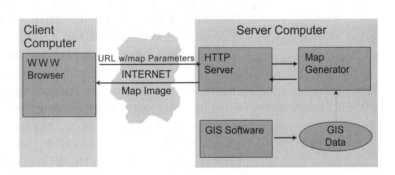

Architecture of a GIS-data map generator.

The fact that this type of engine looks directly at files in the GIS gives it the advantage of always being up-to-date, in that it is reading directly from your current information. The engine does not interact with the GIS software itself; however, with some programming, it can often include much of the functionality of the GIS software.

Because of their ability to work with normal GIS data, and the requirement to program any analysis capabilities yourself, these engines are most commonly used for dynamic map browsers, or for basic queries. For these applications, the products in this category can be quite effective; that is, the processing is faster than a standard GIS because the software is built specifically for a single task.

Feature Manipulation Engine by Safe Software

`http://www.safe.com/fme`

Although this is not a product designed specifically for Internet applications, Feature Manipulation Engine (FME) can work well for certain needs. It is essentially a data format converter (with about 20 GIS and graphics formats currently supported), but it offers many other features, including projection change, simple generalization, and extraction of individual features and themes. In a DGI context, the most common use of FME is to convert data in your GIS to a GIF image, which can then be displayed on a browser, as shown in the following illustration.

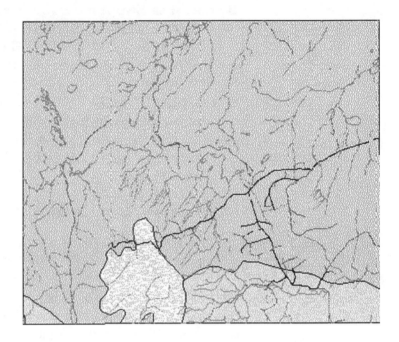

Example map output of FME.

❏ **Form of Software/OS**. The software is a complete program, which can be used either in batch mode (e.g., when used in a Web server), interactively at the command line, or through a customizable GUI. Windows and UNIX 32-bit versions are available.

❏ **Web Server Interface**. Does not contain its own interface to an HTTP server, but it can be incorporated using a CGI script. The script takes the client's parameters and runs FME to generate a map image. This is returned to the server and then to the client.

❏ **Formats Supported**. ArcView Shapefile (ESRI), Generate (ESRI), SDE (ESRI), DGN (Bentley MicroStation), MIF/MID (MapInfo), SAIF, and DXF/DWG (AutoCAD). Several others are under development.

❏ **Output Format**. Outputs color raster GIF images of moderate quality. Can also generate GIS data in most of the formats listed previously. Converters for CGM and VRML are also under development.

ForNet MapServer by the University of Minnesota

`http://www.gis.umn.edu/fornet/docs/MapServer`

One of the few noncommercial products in this chapter, this software was developed for a research contract, but has been made freely available to the public. The server is able to both generate maps and deliver the entire interface. It can also do simple text reports.

The program is controlled through "template files," which define both the map design (including colors, scale, text, and so on) and the HTML interface, which supports layer control, map browsing, and other tools. A ForNet interface and map are shown in the following illustration.

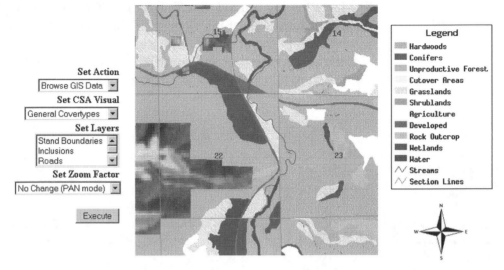

Tile t103109 and Scale 1:26734

[Reset] [Help] [Comments]

An application built using the ForNet MapServer.

☐ **Form of Software/OS**. The software is a complete package, including map generator and CGI scripts. It is distributed as UNIX C source code, which must be compiled on your computer.

☐ **Web Server Interface**. Uses CGI.

☐ **Formats Supported**. Reads directly from ESRI shapefiles (from ArcView).

☐ **Output Format**. Generates GIF images, as well as HTML pages for the interface.

GeoMedia WebMap by Intergraph

`http://www.intergraph.com/iss/geomedia/Webmap`

This system allows for both map display and simple queries, as in the municipal parcel query application shown in the following illustration. The ActiveCGM vector map format allows for a high degree of interaction because each object in the map (e.g., a parcel) can be given a URL for requesting further information. A WebMap application is shown in the following illustration.

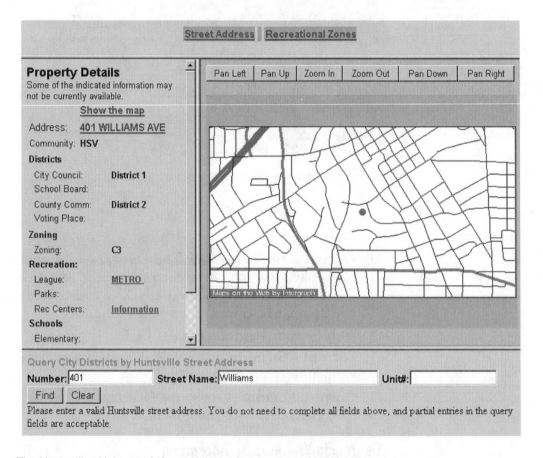

The Huntsville, Alabama, GIS, a public access system using WebMap.

The purchased product is not a turnkey solution because you must design the particular application (consisting of several HTML pages and CGI scripts) yourself. However, Intergraph generally sells consulting time with the actual software to help you with this design process.

❐ **Form of Software/OS**. The purchased product is a set of complete programs for Windows NT, but it requires a considerable amount of programming of CGI scripts to build a complete application.

❐ **Web Server Interface**. Uses CGI to interface with most common Web servers.

❐ **Formats Supported**. Currently supports Intergraph's own MGE and FRAMME software. Support for data formats from other software such as ARC/INFO is either available or under development.

❐ **Output Format**. Outputs vector maps in the ActiveCGM format (an extended version of the standard CGM format, developed by Intergraph). Intergraph offers an ActiveCGM plug-in for Netscape and Microsoft browsers. It is a general-purpose vector graphics format not limited to GIS applications. This format should gain popularity among the GIS community, as well as with other Web users, because it can be used for any type of vector graphics, not just maps.

MapObjects Internet Map Server by ESRI

```
http://www.esri.com/base/products/internetmaps/internetmaps.html
```

This product is an extension to the MapObjects package, a collection of components for building mapping and GIS applications. It can be used to build a wide range of dynamic mapping and GIS applications, using any functionality in MapObjects (which includes a large part of the functionality of ARC/INFO).

For example, the City of Oakland, California (*http://ceda.ci.oakland.ca.us/ps0.htm*), uses MapObjects Internet Map Server to display parcels in the city and to perform simple lookups of parcel attributes. The Oakland Dynamic City Map is shown in the following illustration.

The Oakland Dynamic City Map, built using MapObjects Internet Map Server.

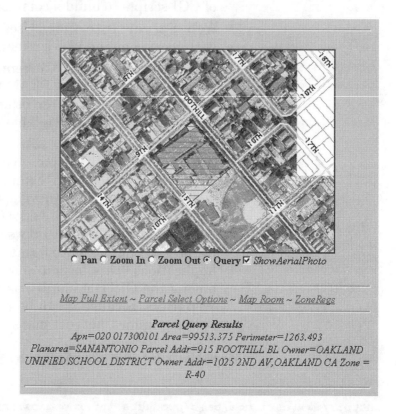

☐ **Form of Software/OS.** IMS is an ActiveX control, which can be incorporated into mapping applications created using the rest of MapObjects in any programming environment that supports ActiveX controls.

❏ **Web Server Interface**. Can use either the NSAPI or the ISAPI interface.

❏ **Formats Supported**. Reads all ESRI formats (shapefiles, coverages, and SDE), as well as images in TIFF, BMP, and other formats.

❏ **Output Format**. Generates either GIF or JPEG images to be used in HTML pages, although you normally pay a licensing fee to Unisys to generate GIF images in your site.

ModelServer Publisher/Continuum by Bentley

```
http://www.bentley.com/modelserver
```

This software is not designed specifically for GIS applications but is geared toward displaying CAD files from Bentley's DGN format, as well as Autodesk's DWG. However, it can be used to display maps from GIS software that uses DGN files as a base, such as Bentley Geographics and Intergraph MGE.

Publisher is essentially a file extractor and converter (similar to FME, previously discussed), taking out specified sections of source files (either spatially or by layer) and converting them to a format that can be displayed on the Web. It is also a scalable solution, automatically starting new copies of the server software. A ModelServer application is shown in the following illustration.

*A map delivered via
Bentley's ModelServer.*

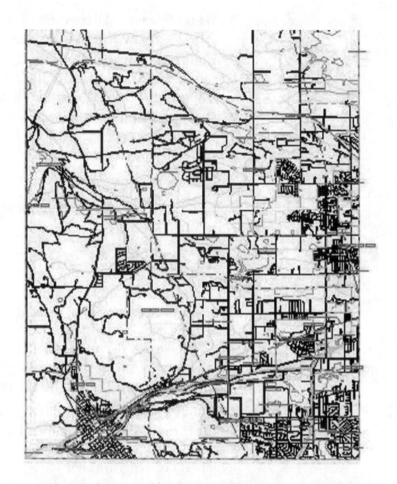

Continuum is a separate product that runs on top of (or rather behind) Publisher. It is an application that stores all data in Oracle's Spatial Data Option (SDO) but performs GIS and CAD analysis using MDL applications such as Bentley Geographics, producing dynamic DGN files to Publisher to deliver over the Web.

❏ **Form of Software/OS**. Complete software, with no GUI interface (administered through Web interface). Available for Windows NT and UNIX.

❏ **Web Server Interface**. Uses NSAPI.

❏ **Formats Supported**. DGN, DWG, and Oracle SDO. Continuum is required in order to recognize GIS attribute data (e.g., from Bentley Geographics) and perform GIS analysis.

❏ **Output Format**. Vector: SVF, CGM, VRML (3D), DGN (2D/3D), and DWG (2D/3D). Raster: PNG and JPEG.

SpatialNet by Object/FX

`http://www.objectfx.com/products/spatlnet.html`

This product is distinctive in many respects. First, both its data structure and application are based on an entirely object-oriented paradigm. Not just a map generator, it is essentially a small GIS, with its own data structures and functions. Because of this, and because the applications are custom built for you, the applications can include almost any functionality you need for Internet and intranet services. In fact, Object/FX software can be used as an entire GIS, with the SpatialNet service as one component. A SpatialNet application is shown in the following illustration.

An application built using Object/FX's SpatialNet.

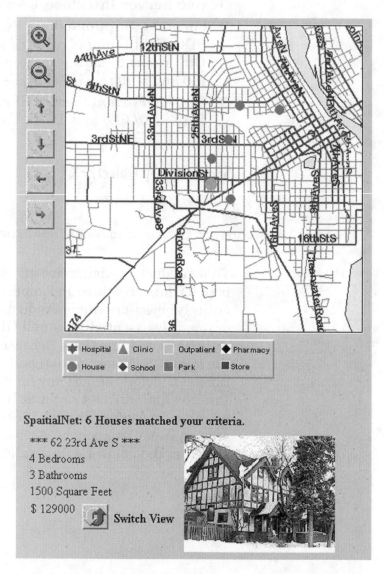

□ **Form of Software/OS.** The product consists of a series of Smalltalk object classes, which build on those of their Visual Companion product. However, it is not an off-the-shelf product but a development package that includes consulting time for Object/FX programmers to develop your custom application.

❒ **Web Server Interface**. Because your application is custom built by Object/FX, it can use any interface mechanism.

❒ **Formats Supported**. The program can store spatial data in any true ODBMS, although Visual Companion (a part of the package) includes a runtime version of the Versant ODBMS. Because no major GIS products use such a system, data will need to be converted.

❒ **Output Format**. Maps are normally created in GIF format. However, other output format types are possible.

Live GIS Interfaces

The products in this category are typically the most powerful and most flexible of products in any of the categories, although they can often be the slowest performers. Here, the DGI program acts as a gateway between the Web server software and a running GIS program, as shown in the following illustration.

Architecture of a live GIS interface.

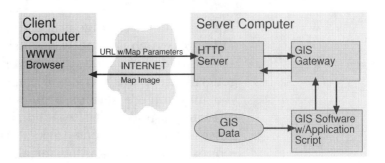

When a request (with parameters) is received from the server, the gateway reformats the request into a set of commands understood by the GIS. These commands are then passed to the GIS program (which is always active, waiting for commands) for processing.

The commands invoke custom-built scripts you have written to perform the tasks for which you have provided Web interfaces (e.g., report the attributes of a selected parcel, or list all parcels within a given radius of the mouse click). This execution of tasks is based on a limited set of parameters provided by the user. The result of the GIS processing, whether in the form of maps or reports, is then sent to the gateway, which passes them on to the Web server, which returns them to the browser for display. (This process is explained in more detail in Chapter 7.)

The primary advantage of this method over the previous method is that you have access to the full functionality of your existing GIS software. If you have considerable resources invested in scripts for automating your user interface, analysis, and plotting, you can make use of these scripts (usually in a modified form) in your DGI site.

The main disadvantage is speed. The fact that the GIS software, which is usually resource hungry, is running at all times may slow the server computer down in general. If several requests come in simultaneously, you may have a serious problem. Because most GIS software is not multithreaded (i.e., the program can process more than one task at the same time), the requests must be handled serially (i.e., one by one).

In the previous two categories, load is usually handled by starting several copies of the generator program as needed. This is effective because the generator program is small and starts almost instantaneously. However, standard GIS software takes several seconds to start, slowing it down even more than the serial processing.

One method suggested for dealing with this problem is to have several server computers running simultaneously, rotating requests among them. Although this solution is robust, providing assurance against slow processing, it may not be cost effective for your organization to purchase and support the necessary hardware.

ArcView Internet Map Server by ESRI

http://www.esri.com/base/products/internetmaps/internetmaps.html

Although the name of this product is similar to the MapObjects product of the same company (previously discussed), it is really an entirely different approach. ArcView Internet Map Server (AIMS) is designed to be an easier program to use—and therefore for a larger market—although it is less flexible than MapObjects IMS because the latter allows you to program any functionality you need.

When AIMS forwards requests to ArcView, an Avenue script is started, which you have written specifically to handle the limited range of applications you wish to support online. This script can be virtually identical to any other Avenue script, except that there is no GUI interface (which will be handled by HTML pages), and maps are drawn to a "virtual view window," which actually results in a raster image returned to the client.

To improve the interface capabilities, ESRI ships AIMS with a Java applet called MapCafé, which you can alter to your needs and deliver with your service (although its use is optional because normal HTML interfaces can also be used). MapCafé, shown in the following illustration, looks almost identical to an Arc-View view window, allowing you to keep uniformity between your existing and online applications.

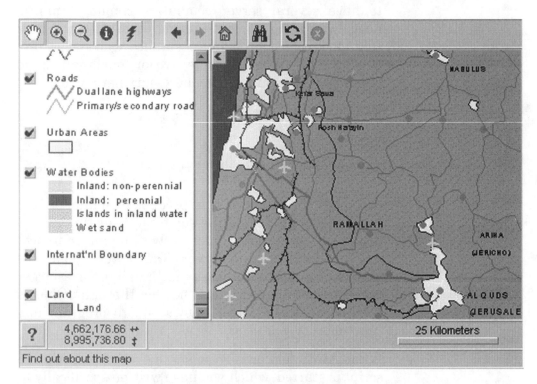

MapCafé, the ArcView-like Java interface that comes with AIMS.

❏ **Form of Software/OS**. The product is actually an extension module for ArcView, not a distinct program. All configuration is done within the copy of ArcView running on the Web server computer.

❏ **Web Server Interface**. Can use either the NSAPI or the ISAPI format.

❏ **GIS Software Supported**. Runs inside a copy of ArcView, and can thus use any data supported by that software.

❑ **Format of Output**. Can create raster images in GIF or JPEG format. However, you are required to pay a licensing fee (to Unisys, not ESRI) if you use GIF in an online application.

GRASSLinks by the University of California

```
http://www.regis.berkeley.edu/grasslinks/about_gl.html
```

The first product to be developed in this (or any other) category, GRASSLinks is still being upgraded by the REGIS group at the University of California, and is in the public domain. As discussed in Chapter 1, several powerful analysis tools—including buffering, overlays, and recoding—are available. A GRASSLinks application is shown in the following illustration.

*An application built
using GRASSLinks.*

□ **Form of Software/OS**. The product is downloaded as source code for UNIX.

□ **Web Server Interface**. Communicates via CGI.

□ **GIS Software Supported**. Operates directly
with the public domain GRASS GIS (*http://
www.cecer.army.mil/grass/GRASS.main.html*).

□ **Format of Output**. Generates GIF images, as
well as associated HTML interface pages.

MapInfo ProServer by MapInfo

`http://www.mapinfo.com/events/prosrv/proserver.html`

MapInfo ProServer is a very robust solution for delivering MapInfo data sets, as well as other data, from databases. This product is not the gateway but is accessed by a CGI script you write. Its responsibility is to pass requests to one or more copies of MapInfo Professional for processing.

This extra level of data delivery allows ProServer to be used in other Intranet applications besides the Web. Once received by MapInfo, the request is processed by a MapBasic script designed specifically for this application, which can access the full functionality of MapInfo. A ProServer application is shown in the following illustration.

An application created using MapInfo ProServer.

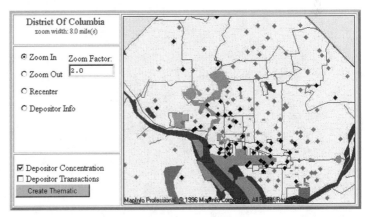

❏ **Form of Software/OS**. Complete package for Windows 95/NT includes the ProServer software, administration tools, and MapInfo Professional.

❏ **Web Server Interface**. Does not directly interface with the Web server but is accessed by custom CGI scripts.

❏ **GIS Software Supported**. Interfaces with MapInfo Professional (in fact, it can work with several concurrently running copies).

❏ **Format of Output**. Generates raster images in GIF format.

Spatial WebBroker by Genasys

http://www.genasys.com/homepage/products/Spatial-Webroker.html

Spatial WebBroker uses a request broker architecture to forward requests from the Web server to one or more copies of Genasys' GIS software. Custom GenaShell scripts then handle the requests to perform the desired functions. A WebBroker application is shown in the following illustration.

An application using Spatial WebBroker.

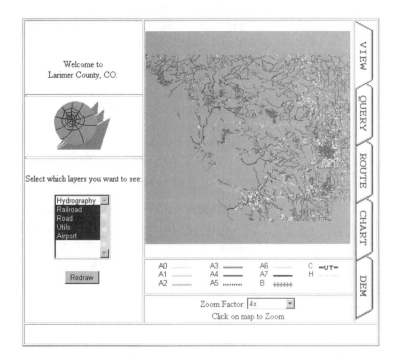

❑ **Form of Software/OS**. The WebBroker is a complete program for UNIX or Windows NT, with no administration interface. Most configuration is done through other Genasys software used for processing.

❑ **Web Server Interface**. CGI.

❑ **GIS Software Supported**. Interfaces with running copies of Genasys' primary GIS platforms: GenaMap, GenaCell, and Henri. Translators are supplied for many other data formats.

❑ **Format of Output**. GIF and JPEG.

Net-savvy GIS Software

This category differs from the others in that this software is on the client side, not the server. The concept of a client-side GIS was discussed in some detail in the previous chapter. The essential difference between a normal GIS program and one that is net-savvy is that the latter is able to incorporate data from across the local network or the Internet as easily as it uses local data sets.

This means that it must be able to read data directly from remote locations, either using distributed file systems such as NFS and dFS or standard Internet/OGC protocols. It must also natively read a variety of file formats, and be able to change projection and coordinate systems on-the-fly.

Although no software adequately meets all of these requirements and cannot be called truly "net savvy," the programs discussed in the following sections are headed in the right direction. The architecture of a net-savvy service is shown in the following illustration.

Architecture of a service based on a net-savvy GIS program.

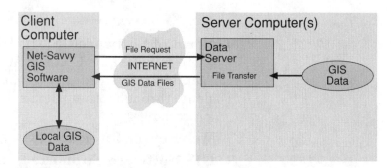

GeoMedia by Intergraph

http://www.intergraph.com/iss/geomedia

Based on Intergraph's Jupiter technology (their first GIS not based on Bentley MicroStation), this software, shown in the following illustration, is designed specifically for conflating data from a variety of sources. The data is delivered by GeoMedia data servers, which are installed on the same machine as the data. Although the GDO transfer mechanism is potentially capable of connecting to remote data over the Internet, the software does not yet support this. Other than this unique capability, GeoMedia is essentially a normal GIS, with a full set of display, analysis, and data editing tools.

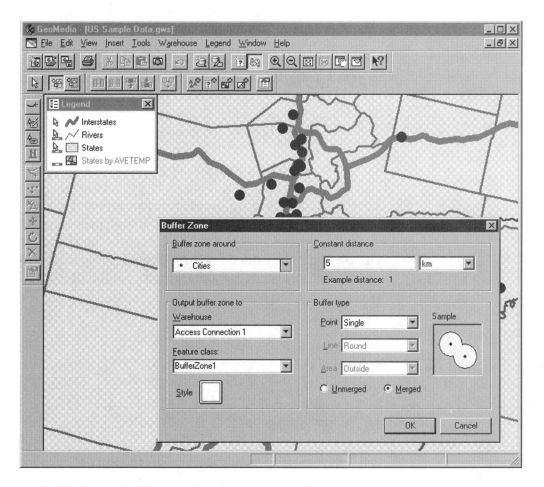

GeoMedia from Intergraph.

❏ **OS Supported**. Available for Windows 95 and NT.

❏ **Internet File Transfer Mechanism**. GeoMedia communicates with its data servers via GDO, a candidate OGC standard discussed in Chapter 3.

❏ **Data Formats Supported**. GDO data servers are available for MGE, FRAMME, ARC/INFO, as well as Oracle SDO, which is not technically a "format." GeoMedia can write to some of these data servers, whereas others have only read-only capabilities.

GRASSLAND by LAS

```
http://www.las.com/grass/LAS
```

This is essentially a commercial-grade version of the GRASS GIS (which is no longer being maintained by the U.S. Army as a public domain product). Its complete analysis and mapping capabilities (including its strengths with raster data) are the same as those for GRASS, with the added feature of having live access to remote data servers, as shown in the following illustration.

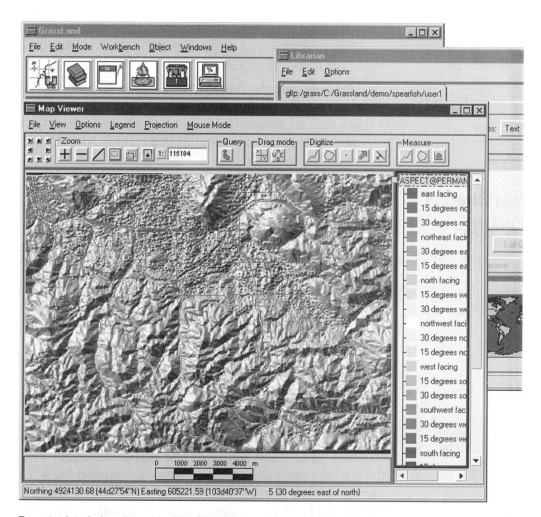

Remote data being viewed in GRASSLAND.

- ❏ **OS Supported**. Available for Windows 95/NT and some UNIX platforms.

- ❏ **Internet File Transfer Mechanism**. Uses the Open Geospatial Datastore Interface (OGDI), a client/server architecture developed by LAS for delivering spatial data over the Internet. This potential standard (it has been introduced to the Open GIS Consortium and ISO) is discussed further in chapters 3 and 8.

❑ **Data Formats Supported**. In addition to the native GRASS format, GRASSLAND supports all sources of data that can be delivered via OGDI data servers, including DIGEST, DXF, ARC/INFO, and Oracle SDO.

Non-Internet Packages

The software packages discussed so far provide specific tools for building DGI applications. However, they are not the only possible solutions. Commercial software and programming tools exist that provide GIS and mapping functionality, which can be used to develop your service, even though they are not specifically designed for the Internet. Two types of these tools are spatially enabled database servers and GIS toolkits.

Spatially Enabled Database Servers

One of the problems in managing and using large GIS systems is that they are proprietary, and have not mixed well with other information in an organization. In the recent past, many of the major database vendors, and some GIS vendors, have introduced tools for incorporating spatial data into a standard relational database. That is, database tables store the geometry of each object, as well as its attributes. Spatial queries can then be run through standard interfaces such as SQL.

As with a normal database management system, custom applications can be built to access this spatial functionality through an application programming interface (API). Therefore, although these extensions are not designed specifically for the Internet, their functionality can be incorporated into Internet applications, provided you are willing to write the program yourself. A possible architecture for such a system is shown in the following illustration.

Architecture of an Internet application built on a spatially enabled database.

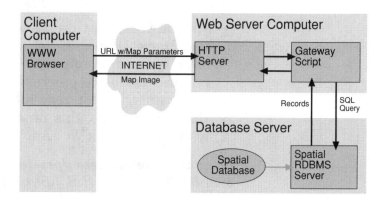

A great advantage of storing GIS data in a DBMS rather than a traditional GIS program, especially when using it for DGI, is scalability. DBMS are designed specifically to work in situations with many users and high load. The software is multithreaded (many requests can be processed simultaneously), is built on a transaction model (which prevents data changes from coming from two directions simultaneously), and has greater security capabilities.

The spatial analysis capabilities of this software are still rather weak, limited primarily to simple spatial queries. However, spatial extensions are available for most of the major database management systems. Descriptions of some of these tools follow.

❑ **Spatial Database Engine by ESRI.** Can actually interface with several database systems to provide spatial functionality. Custom DGI development is not usually necessary to the same degree as the others in this list because it can be accessed through both the MapObjects and Arc-View Internet Map Servers previously discussed.

http://www.esri.com/base/products/sde/sde.html

❑ **Spatial Data Option by Oracle.** One element of the Oracle Universal Server, which adds spatial data types and retrieval methods. Although it requires extensive programming to build a useful application, it is a very robust server. To create DGI applications, this product can be interfaced with Oracle's own Web Application Server.

http://www.oracle.com/products/oracle7/
oracle7.3/html/spatial_opt.html

❑ **Spatial Query Server by Vision International.** Adds spatial functionality to Sybase database servers, with retrieval based on a proprietary spatial extension to SQL. SQS does include some tools for Internet communication.

http://www.autometric.com/AUTO/Vision/
Vis_SQS.html

❑ **Spatial DataBlade by Informix.** Adds spatial functionality to the Informix Universal Server. As an example, CitySearch (*http://www.city-search.com*) has used this product in a DGI service providing location-based searches of tourist and entertainment facilities, including maps.

http://www.informix.com/informix/bussol/iusdb/
databld/dbtech/sheets/sptldb.htm

GIS Toolkits

There are many packages that allow you to build your own spatially enabled applications. If you are willing to program the CGI handling functionality yourself, you can create a map generator like those previously described.

These toolkits are distinct from the scripting interfaces of standard GIS programs such as the Arc Macro Language, MapBasic, MDL, or Avenue because they operate outside the GIS software itself, providing their own spatial functionality (and thus run more efficiently). They may be C libraries, class libraries for an object-oriented language, or even packaged components such as ActiveX Controls (OCX).

The MapObjects package from ESRI might be included here except that it has an Internet interface module and is therefore placed in the GIS-data map generator category. However, the following two products are good examples of this type of tool.

❑ **Hipparchus by Geodyssey Ltd. (*http:// www.geodyssey.com*).** An extensive GIS library based on an ellipsoid Earth coordinate system (rather than the flat map required by most GIS). It is therefore especially useful for global modeling applications, as well as satellite imagery (in fact, it includes special functions for dealing with orbits).

❑ **GeoView by Blue Marble Geographics (*http:// www.bluemarblegeo.com*).** A set of OCX controls for creating maps from a wide variety of data formats. Features include projection and coordinate system changes, and basic spatial queries.

Spatial Catalog Servers

This last category contains capabilities produced by a very different approach from those in the other categories. Programs in this category are used specifically for the metadata search type of service described in Chapter 3. The intent of these services is to assist potential data users in finding the data sets that will best fit their needs. Metadata about these data sets is stored in a database, which users can search or browse to find data sets that meet the criteria of a task.

In a complete clearinghouse service, a dynamic map browser can be included to facilitate location-based queries and to display the coverage of potential matches. You can also provide the capability for users to directly download or purchase the resulting data sets themselves. However, the focus of this type of software is essentially a database, with some simple spatial capabilities.

Spatial Catalog Server Operation

The basic spatial catalog server operation is one in which the user enters a set of desired criteria, and the server does a database query to find matching records. The only difference from a normal database operation is that one of the criteria will often be a desired location (usually a point, rectangle, or other shape).

Handling shape criteria usually requires a spatial object type in the database, and complex spatial query processing. However, point and rectangle queries can be handled by simple arithmetic processes on standard numeric fields (i.e., four fields for a rectangle: east longitude, west longitude, north latitude, and south latitude). Therefore, this type of service can usually be delivered through standard database services such as the following.

❒ **SQL-enabled Web Servers**. These are gateways (similar to the architecture of DGI services) between a Web server and a DBMS, using SQL to issue queries. Most of the major database vendors (such as Borland, Informix, Sybase, and Oracle) have their own HTTP software to fulfill this purpose, and there are many third-party solutions.

❑ **Z39.50 Servers**. This standard is a protocol completely separate from the Web, designed specifically for remote catalog searches (i.e., especially for libraries). This is the means by which FGDC's National Geospatial Data Clearinghouse (*http://fgdclearhs.er.usgs.gov*), described in Chapter 1, is able to integrate many distributed metadata servers. Thus, if you intend for your site to be a part of the NGDC, this is the type of server you must use. FGDC recommends the Isite server software (*http://www.cnidr.org/ir/isite.html*), which provides a complete set of tools for building the service, including a Web interface.

In addition to these solutions there is at least one commercial package specifically designed to build entire clearinghouse sites. This is TerraSoar by Core Software Technology.

TerraSoar by Core Software Technology

```
http://www.coresw.com/TS
```

A complete commercial-grade solution, this software is used by Core Software in its own ImageNet service. The software has a robust means of searching one or more metadata databases, which may be distributed across the network (as is ImageNet). In addition, TerraSoar provides a complete Web user interface, including a global map browser—based on the Digital Chart of the World (a publicly available data set produced by the NIMA)—for criterion entry and coverage display, as shown in the following illustration.

The spatial query interface of TerraSoar.

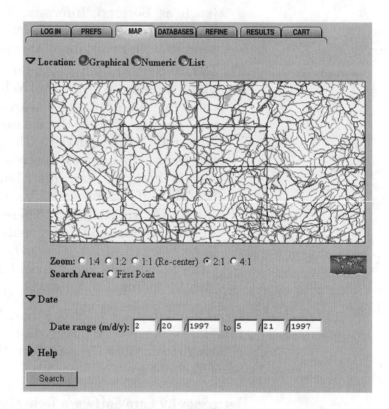

TerraSoar also contains features that make it work well in commercial situations, including security measures, and a "shopping cart" mechanism for purchasing data sets online. It also allows you to register your metadata server with Core Software, making it a member of their ImageNet service (i.e., free advertising). The software is available for UNIX, and is a set of complete programs that interface with your Web server via CGI and require little to no programming.

The Selection Process

The specific characteristics of your situation that factor into the selection process include the intended design of your DGI application, the operating system you wish to use, your financial resources, and whether you have an existing GIS (and if so, the brand of software you use and the format of the data you wish to deliver).

Once you have identified the most likely solutions, you should obtain evaluation licenses of those products (available for almost every one of the packages described in this chapter), and test them yourself. To aid in this process, the following table rates the appropriateness of the types of software with the types of DGI services discussed in Chapter 3. Net-savvy GIS and metadata search/spatial catalog servers are not included because they have a one-to-one relationship between the two chapters.

	Service Type			
Software Type	*Map Browser*	*Data Processor*	*Simple Query*	*GIS Analysis*
Remote Map Generator	Good	No	Limited	No
Proprietary-data Map Generator	Good	No	Limited	No
GIS Data Map Generator	Very Good	Very Good	Good	Good
Live GIS Interface	Good	Very Good	Very Good	Very Good
Spatial RDBMS	Limited	Limited	Very Good	Limited
GIS Toolkit	Good	Good	Good	Good

Summary: Software Options

This chapter has presented software solutions to building online GIS and mapping sites, including remote map generators, proprietary-data map generators, GIS-data map generators, live GIS interfaces, net-savvy GIS software, non-Internet packages, and spatial catalog servers. The options discussed in this chapter are very different from one another. You do not need to make an in-depth evaluation of all of them for a particular situation. Rather, you should assess your needs and then seek matching technologies.

Based on an evaluation of your situation, you should be able to identify the few strategies and software options that are most appropriate. Keep in mind the discussion at the beginning of the chapter of the difficulties in comparing these software packages within and among their categories as you undergo the selection process.

5

Servers and Workers

Allocating Resources for Implementation and Maintenance

It is a foregone conclusion that it will cost money to get your DGI site up and running—and keep it that way. Exactly how much you will need to spend depends on your strategy and the solution you decide to use. As with any other major undertaking, it is vital that these costs be estimated in detail before work begins. Resources will need to be approved and budgeted by management, who need to know the bottom line from the beginning. They are unlikely to allocate large, unexpected sums later in the project.

The costs will arise in computer hardware, software, and labor required to build the service. Each of these resources is described in this chapter. Although the precise costs will vary considerably from one situation to another, you should be able to use this information to develop your own estimates.

Computer Equipment

If you are building a server-side application, you will need a server computer on which to place it. The primary questions to be answered here are "Can I use one of my existing machines as my DGI server?" and "If I need to purchase a new machine or upgrade an existing one, what capabilities do I need?" Other considerations in choosing computer equipment include the following.

❑ Processing power

❑ Where DGI software resides

❑ DGI/GIS interface configuration

❑ Nature of operating system

❑ Nature of Internet connection

Required Processing Power

Whatever DGI software you use (whether you purchase it or write your own), it is inevitable that the program will require considerable computing resources. Although the average Web page requires very little processing by the server (i.e., to find the file on the disk and send it out over the line), there is much processing taking place in even the simplest DGI request. Databases are searched, analyses are performed, and maps are drawn.

This application is extremely load sensitive. That is, if multiple requests are being processed simultaneously (i.e., your site is popular), they must share the same processor, disk, and memory. Therefore, higher load translates directly into slower performance, for not only the DGI processes but everything else running on the computer.

Exactly how much computing power you will need for your service varies. Each commercial software platform will have recommendations for the CPU, memory, and disk space. However, one of the best ways to get to the heart of the matter is to take advantage of others' experiences. The creators of most existing DGI sites described in this book (except perhaps your potential competitors) are willing to discuss their experiences and their setup. Find a site that has employed a strategy similar to the one you are contemplating and ask its producers for advice.

The most important factor to consider when finding a server computer, however, is that it be expandable. As load increases, you will probably want to upgrade the memory, the disk drives, and perhaps even the CPU (or add another CPU).

DGI and Your Web Server

The next issue to look at is where you should put the DGI software. Can it be placed on server machines you already have? In most cases, it will need to reside on the same machine as the HTTP server software that communicates with the Internet, because the two programs are very tightly integrated.

The Web server usually needs to be able to start the DGI program, pass it parameters, and accept results. Although it is technically possible to do this communication between separate computers on a network (using techniques such as RPC, CORBA, DCOM, and IIOP), the current server interface architectures (e.g., CGI, NSAPI, and ISAPI) require the application module to be on the same computer as the HTTP server.

A monolithic server, shown in the following illustration, would house your complete Web site, including mapping applications and regular documents. This type of server is relatively simple to maintain, and works adequately for applications that do not require a great deal of processing or that will not have a large audience.

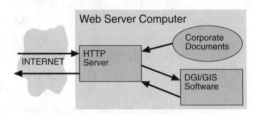

*A monolithic
Web/DGI server.*

However, it is not necessary that the mapping application be on the same server as the rest of your Web site.
Many of the services described in this book employ a
dedicated DGI server computer that is separate from
the computer used for regular pages, as shown in the
following illustration. The DGI machine is accessed
only when maps, or HTML pages that make up the
interface, are requested.

*Dedicated Web
and DGI servers.*

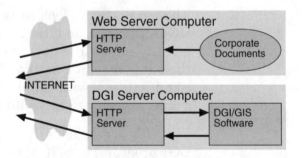

This separation of mapping application and server is
done because GIS and mapping applications require
considerably more processing by the server than the
average HTML page, and it is not desirable to have the
DGI service slowing down the entire Web site. This
method therefore allows the normal documents on
your Web site to be delivered as quickly as they usually are, with the symptoms of heavy load (i.e., slow
response) created by the mapping services restricted
to those services themselves.

The dedicated DGI server must still have its own copy of the HTTP server software, to take requests from the Internet and interface with the DGI program. Therefore, a municipal government, for example, may have two computers, named *www.ci.whoville.ut.us* and *maps.ci.whoville.ut.us*, where a normal document on the Web machine may include a link such as the following:

```
Use our new online GIS to review the proposed site for <a href="http://
maps.ci.whoville.ut.us/map.html?loc=-111.14,38.56&theme=parks"> Cottonwood
Park</a>
```

When someone clicks on the link, they are transferred to the maps machine to browse the GIS. In this case, the dedicated server is generating both the maps and the HTML pages that form the interface. However, in some cases, the latter can be kept on the main Web server, further refining the focus of the DGI server. Maps from one server can be included in HTML pages from another server relatively easily, using the technique described in the previous chapter for remote map generators.

Depending on the load you must support (and hopefully your site will be popular enough to have a high load), it may be necessary to go even further. In the distributed DGI servers approach (shown in the following illustration), there are several machines functioning as dedicated DGI servers, which are virtually identical, probably even sharing a single data source. The user sees no distinction between them, and uses the service as if there were only one server.

Two architectures for distributed DGI servers.

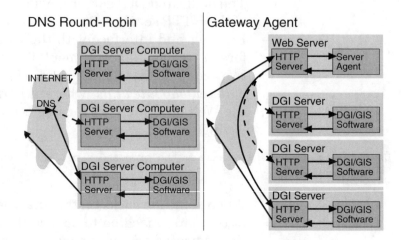

When a request is made, it is automatically routed to one of the machines for processing. The end result is that several $2,000 machines can process many requests in much less time than a single $6,000 machine. The distributed server approach is also scalable, because more machines can be added to the loop to meet increased demand.

There are several techniques available for making the distributed server technique work. The simplest is to use a DNS trick called "round robin," where multiple machines can be given the same hostname, and DNS rotates between them. Therefore, at one time a reference to *maps.xxx.com* may point to one machine, and a second later point to another machine. This is relatively simple if you have access to your own DNS server. However, this configuration does not allocate requests evenly between servers, and some may end up working harder than others.

Another possibility is to have a single front-end gateway server, which receives all requests but does not process them itself. When each request arrives, it determines which of the servers is currently working the least, and forwards the request to the appropriate machine. This is the approach normally taken by Autodesk MapGuide, but it can be done with many of the other products discussed in Chapter 4.

DGI and Your GIS Computers

Depending on your strategy, your DGI software may also need to interface in real time with your GIS software—or at least with your GIS data. There is also a variety of solutions (described in the material that follows) to this side of the equation.

If you are using a live GIS interface (as described in the previous chapter), you will need a copy of the GIS software running on the same computer as your DGI program. This allows the DGI program to make live requests to the GIS program, which is running at all times. (This relationship is shown in the following illustration.)

A monolithic relationship between DGI software and GIS. Hatched area indicates that modules may or may not be on the same machine.

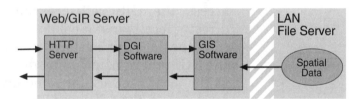

This setup, however, increases the load on the machine because it must support the HTTP server, the DGI program, and the GIS software, all of which may be running simultaneously. In fact, some DGI and GIS software that is not multithreaded must have more than one copy running simultaneously to process more than one request, further increasing the load.

To alleviate this load, you could employ a solution such as MapInfo's ProServer and Genasys' Spatial WebBroker, which are more robust and can connect the Web server to multiple GIS servers over the network, as shown in the following illustration. Note that this is different from the distributed approach previously discussed, because here there is only one DGI module, but the actual processing is passed to several computers and then returned to the DGI module.

A distributed GIS server architecture.

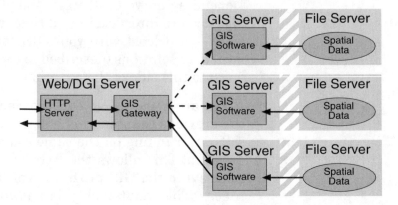

If you are using a map generator, which does not access the GIS software but reads the database directly, locally installed data is not necessary. In this case, the data can be stored anywhere it is readable by the DGI program, including drives on other computers in the local network (or remotely using NFS or dFS), as shown in the following illustration.

A map generator directly accesses data files, which may be distributed.

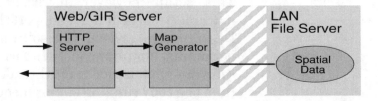

The same is true for the GIS interface solution. Even if the connection is made to the GIS software, and not directly to the data, the GIS can access data from distributed sources.

This distributed-data approach can be very valuable in three situations. If you are combining data from many sources, a single DGI/GIS program can read themes from more than one remote disk. If you are using one (or both) of the two distributed-processing architectures previously described (i.e., distributed DGI servers or distributed GIS servers), all of the distributed processors can share the same data source, making it easier to keep the data source current. The third situation will be the case for most readers. It allows you to use existing GIS data from wherever it is already being stored, which will rarely be the Web server machine.

Although it will rarely get this complex, it is possible to have all three distributed architectures in one system, producing a highly scalable, responsive service, as shown in the following illustration. The drawbacks are the expense for the necessary computers (although many will often be servers and workstations you already have), the management of communications between the machines, and a reduction of speed due to the necessity of transferring large amounts of data over network cables (as opposed to moving it around on a single machine).

A large installation with distributed Web servers, DGI servers, GIS servers, and data servers.

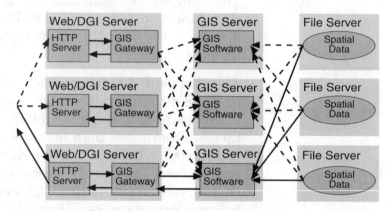

At the opposite extreme, the following illustration shows a very simple overall architecture, with the Web server, map generator, and data all residing on a single computer. This can be adequate for services with one or more of four characteristics: your audience is small (little load), the software is very efficient (each request processes very quickly), there is no GIS data or software in addition to this service, or the computer is very powerful.

A small installation with the Web server, DGI software, and data all residing on a single computer.

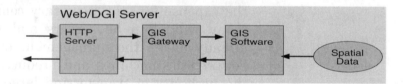

Choice of Operating System

Although you generally have many choices of operating system when purchasing a computer, almost all solutions are designed for either Windows NT Server or UNIX. Therefore, your choice of DGI software will largely dictate which operating system you need to use, or vice versa (i.e., OS selection will restrict your choice of DGI software). This decision will in turn restrict your options for a computer. Some of the solutions can be used in either OS, relaxing your restrictions.

When selecting an operating system (whether before or after your choice of DGI software), there are several factors to consider. UNIX Web servers currently tend to run more efficiently than NT, and are considerably cheaper (the best are free), although the computers are usually more expensive. For the most part, Windows servers are easier to maintain (with GUI administration tools).

It is also important to remember the platform on which your existing GIS software and data reside. Although it is possible to use tools such as the Network File System (NFS) to share information between the two platforms, things work much more easily on homogeneous networks. You also need to consider the expertise and preferences of the staff responsible for maintaining the service, because training costs money.

Internet Connection

Of course, to build an Internet-based DGI service, you will need to be connected to the Internet (which is not necessary for an intranet service). What level of connection should you have? You may already have a connection; if so, you should be aware of the impact a mapping service will have on that line, and perhaps consider upgrading it.

To illustrate the bandwidth necessary to efficiently run a DGI service, consider a hypothetical example. Say you have an interactive map browser, which consists of an HTML page containing a raster map, as well as other graphics and links for navigating the map. An average HTML page is about 1 to 2 kilobytes (K). Maps can vary widely in size, from 5 K to 100 K, but a small raster street map like those in the current map browsers average about 10 K.

The other graphics may vary, but if the interface is consistent, the graphics are reused and do not need to be downloaded again after the first time. Therefore, each time someone clicks on the map or one of the navigation tools, a new HTML page and map are requested, which represents a total of about 12 K.

A frame-relay leased line (the low end of dedicated Internet connections) has a bandwidth of 56 kilobits per second. Therefore, it would take a little over 1.7 seconds (i.e., 12 K x 8 bits per byte ÷ 56 kilobits per second = 1.7 seconds) for your Web server to send the resultant package down this line, although the actual transferred data is more than 12 K because headers for the HTTP and IP protocols are sent as well.

This does not necessarily mean that the user will receive the files in 1.7 seconds, because other factors also play a role, such as the actual distance between you and the user, the bandwidth of the user's connection (which may frequently be a slower dial-up line), and the bandwidth and current traffic of the intervening backbone connections. The 1.7 seconds represents only the time in which the server can push all of the data onto the line.

Of course, the bandwidth of your connection line is not dedicated to this one transfer. If the Web server sends two pages simultaneously (to separate users), they must share the same line. Therefore, each transfer will only be allotted half the bandwidth, and thus take twice as long to transfer. It will be sending other documents as well, to non-DGI users. Also, the Web server is probably not the only user of that line. Personnel who are using e-mail and browsing the Web are also sharing this bandwidth.

A T1 line with a bandwidth of 1.5 Mbps could transfer this one file in less than a tenth of a second. It would take this same line one second to send 15 of the files at once. It is unlikely the load on your DGI service will soon get that heavy, but with the additional traffic from your entire organization, it is not difficult to bog down even this line.

Popular Web sites are most often connected using T3 (45 Mbps, or 30 T1 lines), but this can be very expensive. It is probably best therefore to start out with a T1, as lines can be upgraded if load demands.

Software Purchase

The computer itself is not the only capital resource for which you need to plan. You will also need to purchase the software necessary to make your project work, which can sometimes be as expensive as the hardware.

As described in previous chapters, a DGI application consists of up to three programs that are tightly integrated. The Web server (actually called an HTTP server) receives requests from the Internet or intranet, and returns the maps and other results to the browser. A GIS program may be necessary to do the geographic processing (analysis and mapping), and the DGI program either facilitates communication between the former two modules or performs the geographic processing itself (thus obviating the need for the GIS).

HTTP Server

Whether your DGI server is on the same machine as your main Web server or not, it will need HTTP server software to allow it to communicate over the Internet. If the servers are separate, it is not necessary that the HTTP software used on both machines is the same, although that would make them easier to maintain in terms of learning the software and its performance requirements.

The most important factor to consider when selecting a software is whether it can communicate with the DGI program you have selected. Most of the software described in Chapter 4 use either CGI, NSAPI, or ISAPI as an interface protocol.

CGI is supported on all Web servers, on all platforms, but is slower than the others. NSAPI, although designed for Netscape's line of HTTP servers, is also available on some other platforms. ISAPI is an integral part of Microsoft's IIS server, but has also been incorporated into a few other Web server programs. (In-depth comparisons and recommendations of the available programs can be found in regularly occurring articles in Internet and other types of computer magazines.)

Cost is also a factor to consider. Because it is in the public domain and therefore free, the most popular server software by far is the Apache server on UNIX. The basic version of the Microsoft IIS software is built into Windows NT Server, whereas Netscape and other programs must be purchased, with prices varying widely.

DGI Software

Of the programs reviewed in Chapter 4, only two (For-Net and GRASSLinks) are available without cost. The rest have prices ranging from a few hundred dollars to several thousand, the range of which has more to do with each company's Internet vision than real differences in functionality.

Some see the software as merely a component of the GIS or Web server, and charge very little. Other companies look at their software as an application like any other, and sell it for a moderate application price. Some of the major GIS vendors feel they are essentially giving you an unlimited license to their software (in that any number of Internet users can have access to your application), and believe you are getting a break by only paying for the equivalent of five to ten licenses.

It is expected, however, that the market will even out these pricing schemes in the balance between functionality and cost. In the next few years, prices will likely be more proportional to the functionality of each product.

The other alternative is to build the DGI application yourself. This has been done many times, the TIGER Mapping Service (*http://tiger.census.gov*) being an example. However, what you save in capital costs may be more than offset by labor costs. As an example, the TIGER Mapping Service required over eight man-months of programming. If you decide to use this option, see Chapter 7, which provides guidelines on how to create your application.

Public domain libraries are freely available for drawing images in most of the common graphics formats, but with no up-front purchase of such things as libraries or support services, you will need to develop the rest (parameter handling, reading the data, analysis, and so on) yourself. The format specifications for many of the common GIS data types are published by their producers, but the vendors will charge for any pre-built libraries to read them (such as ESRI's MapObjects).

GIS Software

GIS software may or may not represent an additional cost when building the DGI application. If you are going to use a live (i.e., direct DGI to GIS connection) GIS interface (the only category in Chapter 4 that requires actual GIS software), you most likely already have a GIS installation, for which the DGI server is only a new application. If you currently have no plans for using a GIS outside this DGI application, it is usually best to use a map generator with prepackaged data.

Even if you already have a GIS, there may still be some software costs. The DGI server will usually be dedicated to online service, rather than being one of your general-purpose workstations. Therefore, it may be necessary to purchase an additional copy of the program, or at least another license, to run the software continually on the server machine.

In an intranet situation, there may actually be cost savings in this area. In fact, one of the greatest advantages of using DGI on an intranet is that a large number of personnel can have access to the GIS data and analysis without having to purchase and install individual copies of the full software on every person's desktop.

Development and Maintenance Personnel

In addition to capital expenditures, you will also need staff who can build your DGI applications, then maintain them for the long term. None of the commercial packages described in Chapter 4 are truly turnkey solutions. All of them require some work to tailor the software to your particular application.

Labor is required for many parts of the implementation (described in more detail in Chapter 7). Some of the programs will need only basic configuration, such as the location of your data, and the navigation/query elements you wish to include. Some will require data conversion or other preparation. Others require extensive programming to develop the application.

The major question to be answered here is, can you allot time from your current personnel, or will you need to hire new temporary or permanent staff? The answer will depend on your situation. The sections that follow are intended as guidelines to help you arrive at a solution.

There are three types of skills necessary to the creation of your application. If you are fortunate, the same few people may have all three skills. However, typically with such projects there are numerous people involved, some of whom are only partially involved in the process. This implementation group will probably consist of many of the same people who were in the planning group described in Chapter 2.

For many of the tasks described under the following personnel headings, you will likely be able to allot time from your current personnel, especially if you have a small organization that needs to build just a few, simple DGI applications. However, depending on the amount of work to be done, you may need to hire new staff for certain long-term tasks or maintenance work.

It is also possible to outsource much of this work to professional consultants, especially during implementation, because it eliminates the cost of having to hire and train temporary staff. If you are maintaining the service at your own location, it is considerably more difficult to outsource maintenance tasks.

Designers

One or more people with good graphic design skills should be responsible for designing the appearance and structure of your Web interface (described in more detail in Chapter 6). A well-structured interface to your geographic information will make it considerably more useful than an interface designed by someone who knows the technology very well but has few graphic skills.

The designer will likely create any auxiliary graphics, especially those that are part of your overall organizational image. He or she should also have extensive control over the basic design of your interface documents—including such things as colors, layouts, and fonts—and be conversant with producers of any text.

Your design professional may even have good input on the design of the maps themselves. To be most useful and professional, your maps should not only contain a lot of geographic information but should also be attractive and easy to read. Making them consistent with the rest of your Web site (e.g., in terms of colors and fonts) is also helpful. This is the domain of the graphic designer and the cartographer (or GIS personnel).

Computer Professionals

This group includes people with three types of technical skills. You will need someone with expertise as a system administrator, who can maintain the server hardware and attach it to the rest of your local network and the Internet. This will rarely be someone dedicated to the DGI machine because a single machine should not demand many hours each week to maintain. Your existing computing support staff, if you have such people, should work well for this task.

The second type of person is someone who has experience building and maintaining Web sites (i.e., a webmaster). This person should have expertise in installing and maintaining the HTTP server software, as well as creating HTML pages and simple CGI scripts. Although this task can be quite time consuming (Web servers are still a relatively high-maintenance endeavor), it is not nearly as difficult to learn as other computer technologies. Thus, it is often handled by the same group as the previous task. Together, these two tasks could demand the equivalent of a permanent full-time employee for a sizable Web and DGI site.

The third computer skill is programming, which will often be necessary during implementation, and may be required at times thereafter, when augmenting your services or correcting problems. Less than half of the programs discussed in Chapter 4 are complete programs requiring only basic configuration and interface design. The solutions with the most GIS functionality will require you to build your own custom applications from program tools. As mentioned earlier, if you decide to build your service from scratch, it will require even more programming time.

Depending on the package, you may be developing in the scripting language of the GIS software; in a general-purpose programming language such as C++, Visual Basic, or Java; or creating extensive CGI scripts using a language such as Perl. As opposed to the previous task, programming is something for which extensive experience is very important. Good programmers can develop applications in half the time of novices; therefore, it pays to find experts if at all possible.

Cartographers and GIS Professionals

The third group are those who run the existing GIS installation (if you have one) and are experienced at making maps from it. In implementing and maintaining your DGI services, these people will be responsible for designing the GIS and mapping applications to be placed online. Their geographic and cartographic expertise should ensure that the maps and analyses people obtain from your service will be useful, attractive, and accurate.

Because of their experience in developing GIS applications, they may be the personnel who do at least some of the programming previously discussed. This is especially true if the development is done using the scripting language of the GIS software, in which the GIS professionals likely have some expertise.

Putting the Budget Together

Once you have considered all expenditures, you will need to estimate the total cost and allocate the necessary resources to accomplish the project. This chapter is not a primer on budgeting, with which you are likely familiar, but it should give you some idea of what to expect. The resources previously discussed can be grouped into three separate budgets, depending on when the expenditures will be made. The timing of when the resources discussed in this chapter are needed is summarized in the table that follows.

❏ **Up-front costs**. The items that need to be purchased immediately in order to get started.

❏ **Implementation Costs**. Capital and labor to be expended while your DGI service is initially built.

❏ **Maintenance Costs**. This includes ongoing resources to keep your service running and up to date, as well as for periodically adding new features and improvements.

Resource	Up-front	Implementation	Maintenance
Server Computer	Yes	None	Upgrades
Internet Connection	Yes	None	Upgrades
HTTP Server	Usually	None	Upgrades
GIR Software	Yes	None	Upgrades
GIS Software	Sometimes	None	Upgrades
Designer	None	Part-time	None
System Admin.	Installation	Part-time	Part-time
Webmaster	Installation	Full-time	Part-time

Several things should be noted from this table. First, many of the expenses depend heavily on your situation, including factors such as your DGI strategy, the DGI software you use for your application, and the staff you may already have who can contribute time to the project.

Another important, but often overlooked, expense is the permanent resources necessary to maintain the service. This includes capital expenses to periodically upgrade your server, as well as to fund staff to keep it running and enhance it from time to time. Useful services, especially on the Internet, are never static. You will probably think of new services to include, future releases of hardware and software will facilitate applications that cannot be done today, and, based on user feedback, you will probably find ways to enhance the interface to make it more attractive and easy to use.

Summary: Planning Costs

The most important point made in this chapter is that you need a plan for getting from learning curve to a functional DGI site that does not break the bank. Some of the online GIS and mapping services that can be found on the Internet were developed with very few resources by individuals and nonprofit institutions. Most of these cost a moderate amount to develop. reas However, depending on how sophisticated a site needs to be, it can run into the tens of thousands of dollars. Either way, having a detailed plan should help you foresee most of these expenses, and save money in the long run.

6

Interface Design

Creating a Useful and Attractive Service

The way in which you present your service, and allow users to interact with it, is very important. A well-designed interface will be attractive, functional, easy to understand, and easy to use. On the other hand, a poorly designed site is difficult to navigate to find useful information. Users are not likely to return to your site if it is cumbersome to use, regardless of the mass of information and variety of analysis tools you make available.

The two major elements of almost any DGI service are the map that displays your spatial data and the rest of the interface, which includes buttons and fields to aid navigation of the maps and information retrieval of both spatial and non-spatial information. Each of these involves its own set of considerations in the design process.

Cartography

One of the primary purposes of a map is to represent one or more aspects of the real world. One might think there is only one way of representing "reality." However, there are a wide variety of representations you can make in a map, none of which are perfect, but many of which can adequately communicate any message you wish. To achieve an effective map, there are many aspects of cartographic representation that can be controlled, (i.e., one in which the map reader is asked to see and understand the message you intended).

One of these aspects is the selection of which geographic themes to include. Do you need to show roads, rivers, earthquake hazard zones, the winter range of bighorn sheep, the locations of your offices, or labels for each neighborhood in town? These themes may be represented by point symbols, lines, filled areas, text, or even raster images. For each theme, you have a great deal of control over its symbology, which is the definition of its visual appearance. This definition includes the choice of colors (including hue, saturation, and value), patterns (e.g., area fills, marker symbols, and line styles), line widths, and text styles (e.g., size, typeface, and color).

Thus, your map should be designed specifically for the purpose of your service. There are three primary purposes for maps in DGI services: for reference, for obtaining GIS results, and for accessing general information. Each of these requires its own design. Some design guidelines for each type follow, although it is expected that either your graphic designers or your GIS professionals will have the cartographic skills to produce high-quality maps for your specific purpose.

Reference

This type of map is not intended to be the focus of attention in the service, but is used to show the geographic location of something that is the focus. The primary purpose is to help users relate the location of the object of interest to the real world.

For example, the reference maps used in a data clearinghouse (as described in Chapter 3) have two purposes: to allow users to enter a spatial criterion when searching for data sets, and to show the geographic coverage of candidate data sets that result from the search. The geospatial data set, which is the focus of attention, is not actually shown in the map, except for a simple box representing the area covered by the data set.

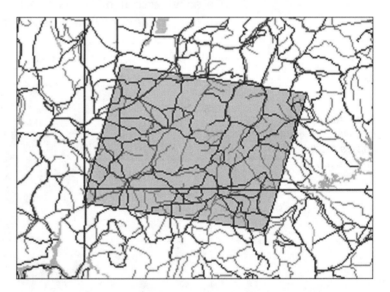

A simple reference map showing the location of an item in a database.

Other examples discussed in this book include continental or national maps showing the office locations of a corporation, which may show nothing but the international and state boundaries. The maps associated with online telephone books, showing the exact location of businesses and residences, are very detailed, but are still used only for reference.

These maps can be very simple in design, showing only those themes—such as roads, rivers, boundaries, and coordinate grids—most helpful in orienting the user. Important landmarks (e.g., highways, mountain peaks, and states) should be labeled, but it is not necessary to label everything on the map. Simple points or shapes that represent the object of interest can then be added (with a symbology that makes them the most easily visible part of the map).

GIS Results

The next type of DGI map also has a very specific purpose. In a service that offers GIS analysis capabilities, the results of a query or analysis will almost always be spatial in nature. Of course, users will want to view these results on a map, as shown in the following illustration.

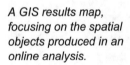

A GIS results map, focusing on the spatial objects produced in an online analysis.

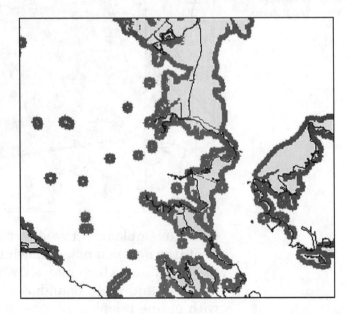

For example, one may use an online GIS interface to ask, "Where are the power lines buried under this piece of property?" The result would be one or more line segments representing the included power lines, obtained by the server's use of an intersection operator between the parcel boundary and the entire power line theme.

This map focuses on the resulting spatial objects; therefore, these should be drawn with a symbology (e.g., color, symbol shape, object size) that makes them stand out from everything else; that is, they are at the top of the visual hierarchy. Slightly lower in the hierarchy, you can include related themes, such as the original objects that factored into the analysis. It should certainly show reference themes (see the previous section) to assist users in recognizing the real-world location of the resulting objects, although these themes can be displayed in the background (e.g., with narrow lines, pale colors, and small labels).

General Information

The third type of map offering is by far the most complex, with no specific predefined purpose for each map. The intent is to show several themes on a map, which the user can browse and interpret for many purposes.

An example of this type of map offering currently popular are street map browsers, including MapQuest (*http://www.map quest.com*), MapBlast! (*http://www.mapblast.com*), Expedia Maps (*http://maps.expedia.com*), and TIGER Mapping Service (*http://tiger.census.gov*). Although street maps are primarily intended for automobile navigation, they are still considered general purpose for several reasons.

First, no single route is displayed (although this type of analysis is possible with some of these services, producing a GIS results map); thus, the same map can be used for determining many routes. Also, other features besides streets are often shown as part of general or detailed information on a map. These features—such as shopping centers, schools, government facilities, parks, golf courses, and airports—further expand the range of activities for which the map can be used.

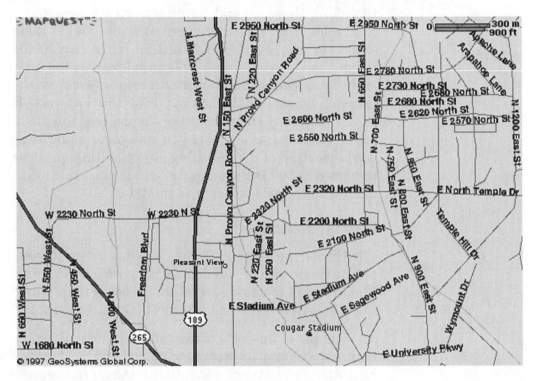

An online street map designed for general information use.

Because there is no single purpose, these maps are usually the most difficult to design. There is no clear hierarchy of importance for the themes because each may be important for different purposes. Also, to support a variety of uses, these maps usually include a large number of themes. One way to solve this dilemma is to use a non-dominant symbology for all of the themes (i.e., narrow lines, small type, subdued colors, and so on); therefore, the map in general does not appear cluttered and the themes do not fight for dominance.

General Design Constraints of Online Maps

Regardless of the types of maps included in a service, the Web medium provides some unique design considerations cartographers and GIS professionals may not be accustomed to handling. First, the size of each map is very limited. Your map will need to fit on the screen, inside the Web browser. If it is larger than the available space, users will have to scroll to see all of it. This makes the map less attractive and much less interactive (i.e., it will take users more time to see parts of the map).

The space available depends on many factors. On a standard 14-inch monitor, the screen size is generally 10.5 inches by 7.5 inches. The Web browser itself uses some of that space (for scroll bars, status bars, button bars, menu bars, and so on). If the browser is maximized (using the entire screen), the document window would be about 10 inches by 6.5 inches.

In addition to the map, you will have other items in the HTML document, such as navigation buttons, some of which will need to be visible onscreen at the same time as the map. This may leave a maximum map size of 9 inches by 5 inches, or much smaller if you have a complex interface. Obviously, if your users have larger monitors, they will have more screen space available, but in the eclectic world of the Web, you should usually design for the lowest common denominator. Thus, the area your map can cover at one time is limited.

Screen resolution (i.e., the size of the pixels) is another limiting factor. For a desktop printer, 300 dots per inch (dpi) is a minimum resolution, and high-quality printed matter usually has a resolution of 2,540 dpi or more. This high resolution allows for more readable text (especially at small sizes) and smooth fills (i.e., the individual dots are not as visible). However, monitors generally have resolutions between 60 dpi (640 pixels x 480 pixels on a 14-inch monitor) to 100 dpi (1,280 pixels x 1,024 pixels on a 17-inch monitor). This coarse resolution limits the minimum size of text, the geographic detail you can include, and the appearance of patterns.

Combined with the physical screen size, this resolution yields a general maximum size for raster maps of 550 pixels by 300 pixels. If the map is larger than this, many of your users will need to scroll horizontally or vertically to see the rest of the map and associated interface, and the service will not be as easy to use.

One advantage of onscreen maps over printed maps is in the area of color. Generally, it is very easy and economical to use many colors onscreen, whereas it incurs additional expense on printed media. In some cases, color can be used to overcome some of the limitations of resolution in the same way that a television uses an almost infinite variety of colors to produce realistic images with a resolution worse than a computer monitor.

Another possible advantage of online maps over traditional media is the ability to use animation and surface models for representing time-based and 3D geographic information. Animation can be done through either GIF image rotation—a technique whereby a series of GIF images can be shown one after another—or through the use of an animation-capable browser plug-in, such as Macromedia ShockWave (*http://www.mac-*

romedia.com / shockwave). Three-dimensional models can be delivered using the standard Virtual Reality Modeling Language (VRML), which can be automatically generated from GIS data using Bentley's ModelServer Publisher or other software described in Chapter 4.

One more consideration is that the maps are usually generated automatically, without the intervention of a professional cartographer. Most of the representation in a map can be automated quite easily, such as the drawing of points, lines, and areas. However, other techniques are more difficult, such as the placement of text, and some graphics tricks can only be done manually. The general rule here is that because these maps must be done manually they should be kept relatively simple, so that these fancy techniques are not necessary.

Web Interface

In almost every type of DGI application, the map by itself is not very useful. Any complete service must include other tools that allow users to access the full functionality of your GIS and your spatial data. The map as well as these other interface elements are encoded in an HTML document, which is displayed in the browser window, or as a part of a Java or ActiveX applet or browser plug-in.

The interface of any DGI service may include interface elements that perform three types of functions: navigating or browsing the map, controlling the map's appearance, and performing GIS queries or analyses. The following illustration shows how the parts of the interface are arranged in a real application, the TIGER Mapping Service (*http: / / tiger.cen sus.gov*).

A full-featured map
browser interface used
in the TIGER Mapping
Service.

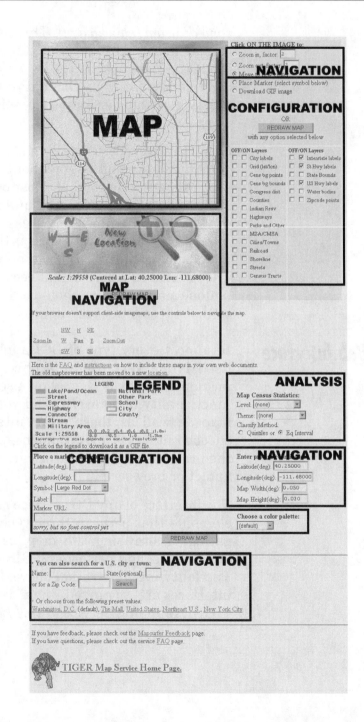

All of these tools are displayed using one of four visual methods: graphic icons (e.g., an "up" arrow to indicate movement north), text links (e.g., the word *North* moves to the north), directly pointing to the map image (e.g., clicking in the upper part of the map to move north), or through text entry blanks (e.g., typing in a larger value for "Latitude of Center" to move north). Several examples of three of these methods are shown in the following illustration.

Types of design elements.

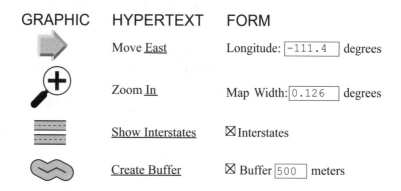

GRAPHIC	HYPERTEXT	FORM
	Move <u>East</u>	Longitude: `-111.4` degrees
	Zoom <u>In</u>	Map Width: `0.126` degrees
	<u>Show Interstates</u>	☒ Interstates
	<u>Create Buffer</u>	☒ Buffer `500` meters

The first two methods (graphic and hypertext) are implemented using normal HTML links (i.e., <a> tags), whereas the last one (form) requires the use of HTML forms. If you use forms, you will need to remember to include a means of sending the entered information to the server. Clicking on the map usually submits the form immediately. Alternatively, you can include a "submit" button.

Clickable raster maps (the third type of element) can be created using a special type of HTML link, or as part of a form. The first option uses the following syntax:

```
<a href=XXX><img src=YYY ISMAP></a>
```

This makes the entire image a link to the URL XXX. However, the ISMAP option of the <a> tag tells the server to not only request the XXX page but to send it the coordinates (in pixels) of the place the user clicked on the image.

This technique is commonly used in Web pages that contain graphic menus (i.e., all of the options are drawn in a single image). The appropriate XXX URL interprets the mouse click as falling within the area covered by a particular menu option and forwards the user to the appropriate page.

To include a map as part of a form, use the <input type=image> tag. When a user clicks on the image, the entire form is immediately submitted, including the coordinates of the mouse click, as in the previous option.

The trick is in the script (i.e., XXX) requested by either of these techniques. In the case of a map, the image does not represent buttons with set extents, but positions on the earth. The pixel coordinates representing where the user clicked are useless to the DGI program. What is needed are the real-world coordinates as represented in the underlying GIS data.

To obtain these coordinates, the submitted point is "unprojected." Because a projection algorithm with set parameters was originally used to create the map image (with features drawn in pixel coordinates) from the database (which uses real-world coordinates), a reverse algorithm can be run, with the same parameters, that converts the pixel coordinates of the mouse click back into real-world coordinates. These real-world coordinates can then be used by the DGI program to perform whatever function was requested.

Implementation of the three general types of tools is described in the material that follows, along with general design guidelines. A fairly complete set of options follows, but for your specific application, you will rarely wish to use all of these interface elements. You should select only those tools you need, then design the overall interface to include them. Other tools can always be added later if the need arises.

Map Navigation

In many types of DGI applications, users have the ability to browse maps. This means that the map being displayed covers only a portion of the area covered by the underlying spatial data, and users can move that "window" to an area of interest. This may involve moving to a specific location (e.g., a city or a latitude/longitude); panning, or moving in a certain direction (north, south, east, west, and so on) from the current location; and zooming in (i.e., increasing the scale, making objects on the map larger) or out (i.e., decreasing the scale, allowing the map to cover more area).

If the map is in a vector format, it will usually be displayed using a Java applet or browser plug-in (such as MapGuide from Autodesk, as shown in the following illustration, or ActiveCGM from Intergraph), which has its own browsing tools. However, if the maps are in GIF or another raster format, you will need to design your own browsing interface, either in HTML or in Java or ActiveX (e.g., ESRI's MapCafé or MapQuest).

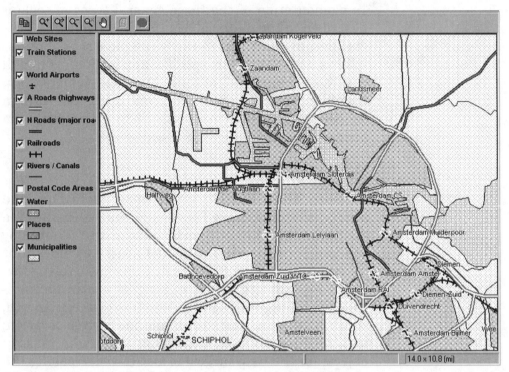

Map navigation tools built into the MapGuide browser plug-in (upper left corner).

The HTML solution, currently most common, used in such sites as the TMS (previously shown), are described in more detail throughout this chapter. The Java/ActiveX approach uses roughly the same ideas, but the code to accomplish it is different. There is a variety of possible designs for the navigation tools, but the manner in which they operate varies little.

When the user's Web browser requests the DGI application (either an HTML page or a CGI/ISAPI/NSAPI script that generates an HTML page), it includes a set of parameters in the request (often given in the URL) that defines the desired map location. These parameters include the latitude and longitude of the map's center, the width and height of the geographic area to be included (or a scale factor), and the width and height of the map image in pixels. An example URL for the page might be:

```
http://www.xxx.com/Webgis?lg=x1&lt=y1&mw=x2&mh=y2&iw=x3&ih=y3
```

Here, *Webgis* is the name of the HTML page (or a CGI script that generates it), *x1* and *y1* are the center longitude and latitude (in degrees), *x2* and *y2* are the width and height of the geographic area to be covered (in degrees), and *x3* and *y3* are the width and height of the map image (in pixels).

These parameters are passed on to the generator, which creates the map image. The parameters are also used by the navigation tools. Each of the interface items is actually a hyperlink, which points to the same URL as the page currently displayed, but with slightly different parameters, as shown in the following illustration. For example, to pan to the east, the page would be requested with parameters identical to the currently displayed page, except with a slightly increased center longitude. To zoom out, the page would be requested with a larger geographic width and height.

Called with URL: http://www.xxx.com/webgis?lg=x1&lt=y1&mw=x2&mh=y2&iw=x3&ih=y3
webgis script calculates the following parameters:
x1l = x1 - x2   x1r = x1 + x2   y1u = y1 + y2   y1d = y1 - y2
x2i = x2 / 2   x2o = x2 * 2   y2i = y2 / 2   y2o = y2 * 2
webgis script generates the following HTML code:

```
<p align=center>
<a href="http://www.xxx.com/webgis?lg=x1&lt=y1u&mw=x2&mh=y2&iw=x3&ih=y3"><img src="up.gif"></a><br>

<a href="http://www.xxx.com/webgis?lg=x1l&lt=y1&mw=x2&mh=y2&iw=x3&ih=y3"><img src="left.gif"></a>

<img src="http://www.xxx.com/webmap?lg=x1&lt=y1&mw=x2&mh=yw&iw=x3&ih=y3">

<a href="http://www.xxx.com/webgis?lg=x1r&lt=y1&mw=x2&mh=y2&iw=x3&ih=y3"><img src="right.gif"></a><br>

<a href="http://www.xxx.com/webgis?lg=x1&lt=y1d&mw=x2&mh=y2&iw=x3&ih=y3"><img src="down.gif"></a><br>

<a href="http://www.xxx.com/webgis?lg=x1&lt=y1&mw=x2i&mh=y2i&iw=x3&ih=y3"><img src="zoomin.gif"></a>

<a href="http://www.xxx.com/webgis?lg=x1&lt=y1&mw=x2o&mh=y2o&iw=x3&ih=y3"><img src="zoomout.gif"></a>
```

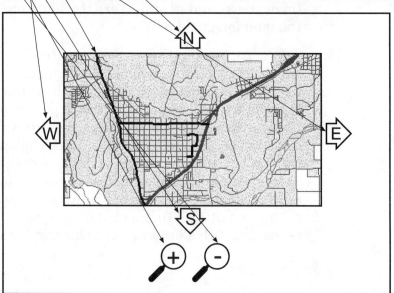

Map navigation tools implemented in HTML.

Panning

Panning tools can be displayed in several ways. In the following illustration (Ameritech Yellow Pages, *http:// /yp.ameritech.com*), graphical arrows are placed surrounding the map. To move in a certain direction, the user clicks on the appropriate arrow. Text links such as "East," "West," and so on can also be positioned in this manner.

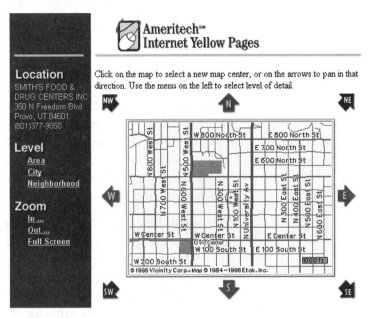

Panning icons surrounding a map in the Ameritech Yellow Pages.

The alternative is to use a compass rose, as shown previously in the TMS illustration (below the map and to the left). This is a small graphic placed near the map, with arrows for the various directions. Users click on the arrow pointing in the direction they wish to pan.

Another possible solution is to use a recenter feature. When the user clicks somewhere on the map image, a new map is generated, with the requested point as the new center. This is accomplished with one of the two clickable graphics techniques discussed earlier in this section. The pixel coordinates of the place the user clicked on the image are sent to the server, where they are un-projected into a latitude and a longitude. These new coordinates then become the new map-center parameters for the HTML page and the map.

The recenter tool works well for fine tuning the map location, whereas the other tools are better for moving greater distances. Thus, many map browsing services, such as the TMS, include both a recenter feature and one of the former two styles of directional tools.

Zooming

There are two common methods used in Web interfaces to allow users to increase or decrease the scale of a map. The first is to include a pair of graphic or text links near the map (e.g., the magnifying glasses in the TMS illustration); one to zoom in by a prescribed amount (e.g., double the scale), the other to zoom out.

The second method is to allow users to click on the map itself to zoom in or out on a specific location. This combines the zoom function with the recenter function. Because all three navigation functions (recenter, zoom in, zoom out) involve clicking on the map, it is common to include checkboxes (in the TMS example, to the right of the map) to allow users to select which function will be performed when they subsequently point.

With either method, the zoom factor (e.g., multiply the scale by two or by five) can be fixed, or the user can be allowed to specify the factor. The latter is accomplished with text entry fields, created using the forms feature of HTML, as shown in the TMS illustration.

Absolute Positioning

There are two common methods for allowing users to move directly to a specific location. The first is to include text entry fields for the latitude and longitude of the center of the map, as well as the other map parameters (shown in the lower right of the TMS interface). When the user enters these and submits the form, the DGI page is requested using the entered parameters instead of the current ones. This is very easy to implement at the server, but users will rarely know the exact coordinates of their area of interest.

The other method is to search for a geographic feature with which the user is familiar, such as a city, a river, a county, or even a street address. The server either searches a gazetteer or an address database to find the desired feature, determining its latitude and longitude from the database and adjusting the map parameters accordingly. The gazetteer tool is shown at the bottom of the TMS interface. Alternatively, MapQuest (*http://www.mapquest.com*), for example, employs an address location form, as shown in the following illustration.

The address location form in MapQuest.

Map Appearance Configuration

You may also allow users to have control over the general appearance of the map, as well as the geographic location covered. This can range from including a small selection of color schemes to having almost complete control over the map.

Depending on the DGI software you are using, there are many controls that can be made available. These include which layers should be displayed (e.g., roads, cities, rivers, and statistical data), placing markers on the map at prescribed locations (e.g., your store location), and changing the symbology (colors, line styles, fill patterns, and so on) to be used.

These are implemented in the same manner as navigation tools. The appearance of the map is controlled through a set of parameters, often given in the URL. For example, the hypothetical URL for the map generator previously discussed (under "Map Navigation") could also have the following parameters:

```
http://www.xxx.com/Webgis?. . . &on=a,b,c&mln=x4&mlt=y4&ml=label&pal=p
```

where *a,b,c* is a list of themes to be displayed, *x4* and *y4* are the longitude and latitude of a marker to be added, *label* is the text to place next to the marker, and *p* is the name of a color palette to be used.

Again, the visual items in the HTML page (e.g., graphical buttons, text, and form fields) are links to the same page address, but with the parameters changed appropriately. The three configuration functions can be implemented in a variety of ways, as described in the sections that follow.

Theme Display

Because any map represents only a small selection of the full complexity of the real world, the data themes you choose to display on a map can vary widely. The map may include themes for reference (e.g., roads, administrative boundaries, and latitude/longitude grids), themes relating directly to the map purpose (e.g., vegetation zones, study area boundaries, statistical data, and points of interest), themes generated as a result of the user's GIS analysis request, and labeling for any of the foregoing. In the TMS service, there are over 20 layers that can be displayed, whereas a more simple street map browser, such as MapQuest, usually has fewer than ten.

Obviously, including all of these 20 themes in a single map is not possible, and if it were, the map would not be readable. It is generally advisable to have no more than seven to eight separate themes displayed at once. For example, the TMS by default has six layers displayed: county/state boundaries, cities, public lands (parks, military, airports, and so on), water features, roads, and city labels.

Therefore, in many cases, a few themes need to be selected from the available set. If your maps have a specific purpose (i.e., the "GIS results" or "reference" types of maps described in the first section), you will likely decide which layers should be included to best meet this purpose. However, for a general information map, you should allow users to control which ones to include, depending on their own purposes.

There are two ways of implementing this control as map parameters. One way is to have a parameter (e.g., ...&layers=road,water,parks) that lists all features to be displayed. The other approach is used in TMS. A configuration file on the server defines the default themes to be displayed at the given scale; the parameters entered by the user define deviations from these defaults (i.e., turn layers on that are normally off, and turn layers off that are normally on).

Both of these types of control are usually accomplished through form fields. The form can include either a series of checkboxes for each possible theme, as shown in the TMS interface (in the center right), or a listbox, as used in ForNet ForestView (*http://www.gis.umn.edu/fornet/gds/forestview*), shown as the second and third items at left in the following illustration.

The ForNet Map Server interface includes control over the layers displayed.

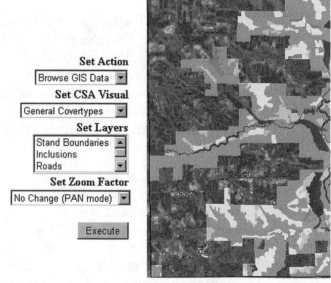

Tile t108109 and Scale.

Markers

This is the most commonly available tool for customizing DGI maps in map browsers, because it has a wide variety of applications. In some cases, you will want to display your own selection of places on the map, such as offices, stores, tourism sites, and homes for sale; for other applications, you may wish to allow users to enter their own locations to be displayed.

These locations are defined by their latitude and longitude, or whatever coordinates are used in the underlying GIS data. These coordinates can be entered directly, as in the lower left corner of the TMS interface, or derived from a database (e.g., a gazetteer, street address table, or telephone book), as is the case with MapQuest, which places a marker at the desired location as well as moving the map there.

In the latter case, you would include form fields to allow users to search the database for such things as a feature or a business name, address, or city. When the user submits the form, the server searches the database, which includes not only the appropriate key fields (address, name, and so on) but a latitude and longitude for each record.

Depending on the database and the ambiguity of the entered keywords, there may be several matches. If this is possible in your application, you will need to include a method for displaying the candidate records to the user, so that they can select which one they wish to display. This is used in most of the online telephone books, such as the Yahoo! Yellow Pages shown in the following illustration.

YAH**OO!** YELLOW PAGES   DATA BY Lookup**USA**

Search Address : Provo, UT

Business Name	Address	Phone	Directions	Maps	
Kent's Market	1209 N 900 E, Provo, UT 84604	801-375-9452	DriveIt!	MapIt!	☐
Oak Hills Gas N Stuff	1220 N 900 E, Provo, UT 84604	801-375-5771	DriveIt!	MapIt!	☐
Cougar Conoco	818 N 700 E, Provo, UT 84606	801-374-1077	DriveIt!	MapIt!	☐
Circle K	1240 N University Ave, Provo, UT 84604	801-373-7606	DriveIt!	MapIt!	☐
Hart's Gas & Food	1429 N 150 E, Provo, UT 84604	801-375-2477	DriveIt!	MapIt!	☐
7-Eleven Food Store	496 N University Ave, Provo, UT 84601	801-373-7267	DriveIt!	MapIt!	☐
Gas-N-Go	47 E 300 N, Provo, UT 84606	801-375-1412	DriveIt!	MapIt!	☐
Storehouse Markets Savings	630 N 200 W, Provo, UT 84601	801-375-8962	DriveIt!	MapIt!	☐

[MORE Listings]   MapIt!

Another method for placing markers is to allow users
to point on the map directly to the desired location.
This option would possibly need to be combined with
the recenter and zoom actions previously described,
using radio buttons as shown in the TMS interface
(upper right in the previous illustration).

In addition to the location of the marker, further customization may be made available. Users may want to
use various graphical symbols (e.g., dots, crosses,
arrows, or pins) to mark the location. You would need
to design several symbols into the interface, which
users would access from a list in the form.

You might also make it possible for users to place a
text label next to the marker, possibly even with a
choice of font and size. This text is usually entered into
a form field (as shown in the lower left of the TMS
interface), and both parameters are passed along with
the rest of the form.

If you are producing maps using your own points of interest, your map generator can read these four basic parameters for each point (latitude, longitude, symbol, label) directly from your database or a file, but if you are plotting markers for your remote users, you must have a means for them to submit the parameters to your server. Including them in the URL works for one or two markers, and this technique is used in TMS, the Xerox Map Server, MapBlast!, and other services. However, placing many points would produce a URL far too long and cumbersome.

The alternative is to send the desired information another way. Whereas the GET method (a type of request in the HTTP protocol) must have all parameters in the URL, the POST and PUT methods can send information attached to the URL, but not part of it. Thus, they are commonly used for forms because they can send larger packets of information. Either of these methods can be used to allow users (or your own DGI application) to submit a larger list of points, entered using a textbox in a form or even from a disk file.

Symbology

The last type of control that can be made available to users is in the symbology of the map. Because you are spending the time to design maps that will be effective for the purposes of your service, you will rarely want users to have complete control over their appearance. However, some degree of customizable appearance can be a help to users who will be creating maps for their own purposes, as well as to allow them to make your service more appealing to their own tastes.

For example, Vicinity's MapBlast!, as shown in the following illustration, has four overall color schemes available to users. Each scheme, designed by Vicinity, defines a slightly different symbology, primarily the colors of the background and the streets. Users can select any of the four icons to the left of the map to change the color scheme.

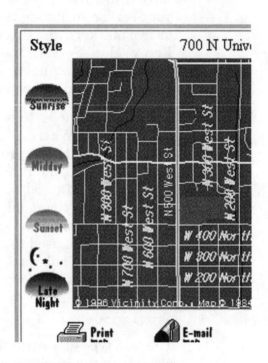

GIS Analysis and Query

The fourth part of the interface of a DGI application has a wider variety of options than the rest because it is completely dependent on the purposes of your service. If you wish to give users access not only to your spatial data but to the analysis capabilities of your GIS software, you will need to devote part of the Web interface to these functions.

The following two illustrations show how analysis tools may be incorporated into DGI services along with the other elements. The first illustration shows the interface for the Kansas City GIS (*http://maps.intergraph.com/kcWeb*), built with Intergraph GeoMedia Web-Map. The second illustration shows the San Francisco Bay/Delta GIS interface (*http://regis.berkeley.edu/grasslinks*), built using GRASSLinks.

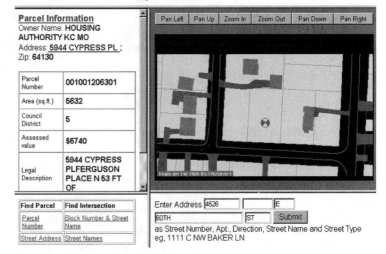

The interface of the Kansas City GIS, which offers GIS query capability and map browsing.

The REGIS San Francisco Bay/Delta GIS includes a wide variety of analysis capabilities.

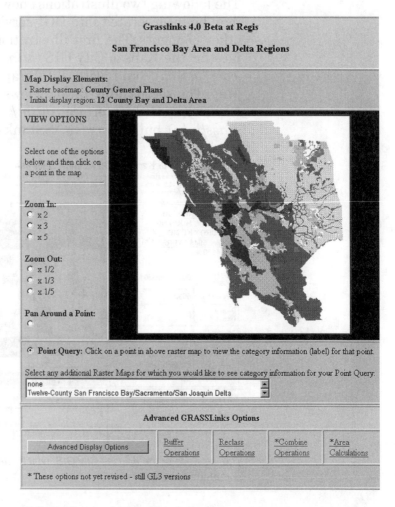

Grasslinks 4.0 Beta at Regis

San Francisco Bay Area and Delta Regions

Map Display Elements:
· Raster basemap: **County General Plans**
· Initial display region: **12 County Bay and Delta Area**

VIEW OPTIONS

Select one of the options below and then click on a point in the map

Zoom In:
○ x 2
○ x 3
○ x 5

Zoom Out:
○ x 1/2
○ x 1/3
○ x 1/5

Pan Around a Point:
○

○ **Point Query:** Click on a point in above raster map to view the category information (label) for that point.

Select any additional Raster Maps for which you would like to see category information for your Point Query:
none
Twelve-County San Francisco Bay/Sacramento/San Joaquin Delta

Advanced GRASSLinks Options

| Advanced Display Options | Buffer Operations | Reclass Operations | *Combine Operations | *Area Calculations |

* These options not yet revised - still GL3 versions

Although you are allowing users to access operations in your GIS software, you cannot (and do not want to) have them use the software directly through the same interface with which you use it. Rather than the standard command line prompts or GUI window systems you can use locally, the Web has its own interface, which you must use. In almost all cases, you will not need to replicate the entire GIS interface, with all of its functionality, on the Web. Instead, there will be a few specific applications duplicated from your standard GIS to the DGI service.

Attribute Queries

Most of the GIS applications currently seen in DGI services involve rather simple searches. The method is essentially the same as a search in any database, except that the records are spatial features to be displayed on a map (perhaps along with a normal text report).

The user enters desired keywords or other criteria into an HTML form, and submits it to the server. The DGI software then searches the appropriate GIS data theme for objects that have attributes matching the criteria. A new map is then generated in which the selected features are highlighted. Queries of this type include the following:

❏ Which parcels are owned by Bill Jones? (In this case, the theme is *PARCELS*, and the criteria is *OWNER="Bill Jones".*)

❏ Which census tracts are more than 40 percent Hispanic? (The theme is *TRACTS*, and the criteria is *PERCENT_ HISPANIC>0.40.*)

❏ Which houses for sale have a price between $115,000 and $150,000, and at least four bedrooms? (The theme is *HOUSES*, and the criteria is *PRICE>=115000 and PRICE<=150000 and BED>=4.*)

A rather simple HTML interface to this search function would include two text entry fields: one for the search theme, the other for the criteria. Although this would be very flexible, it would not likely be useful for an audience unfamiliar with the technical details (i.e., exact names of the themes and the fields available) of your

GIS data. For applications that require this degree of flexibility (e.g., in an intranet), you should at least include one or more pages of documentation describing the available options and required syntax, with a link to this documentation from the search form.

A more constrained, but easier to use, interface is used in sites such as the Kansas City GIS previously discussed (shown in the lower portion of the screen). Only one theme is available for searching at any given time. If you want to make it possible for several themes to be searched, you can create a separate HTML form for each theme. Provide a list of searchable themes, so when a user selects a theme, the appropriate form is displayed.

You then include text entry fields for each attribute field in the data theme. This makes it clear exactly what users are able to do, even if their options are limited. An example HTML fragment for this type of form and the displayed form are shown in the following illustration.

An HTML form for performing an attribute query on a GIS.

```
<form action="http://www.xxx.com/webquery" method=POST>
Enter your search criteria by clicking on the fields you want to search and entering text. Then click
on "Search."<p>

<input name=owner_cb type=checkbox> Owner's Name: <input name=owner size=40> (last, first)<br>

<input name=address_cb type=checkbox> Address of Property: <input name=address size=50><br>

<input name=value_cb type=checkbox> Assessed Value:
<select name=value_op><option>= <option> &lt; <option> &gt; <option> &lt;&gt; </select>
<input name=value><br>

<input type=submit value="Search">
</form>
```

Enter your search criteria by clicking on the fields you want to search and entering text.   Then click on "Search."

☐ Owner's Name: [   ] (last, first)
☐ Address of Property: [   ]
☐ Assessed   Value [▼] [   ]

[Search]   =
              <
              >
              <>

Whether you use either interface, or something in between, the process on the server end is the same. The DGI program must reformat the received criteria into a syntax that can be understood by your GIS software, to which the query is then submitted for processing.

Map-based Spatial Queries

The second type of GIS tool that can be included in the interface is a query in which the location is one of the criteria. Because the entire application is spatial in nature, this is a very common task. Common queries might include the following.

❑ Identifying a feature displayed on the map and/or requesting its thematic attributes. For example, "What is the name of *this* street?" (where the word in italic refers to a location indicated by the user on the map with the mouse).

❑ Defining an *area of interest* to be submitted as a criterion for a search for features of a particular theme, often along with attribute criteria, as described in the previous section. For example, "Which parcels in *this* area are owned by Bill Jones?"

❑ Selecting one or more features for use in one of the more complex analyses (described in the material that follows.) For example, "Generate a 50-meter buffer around *this* street and *this* street."

Except for very simple queries, which can be handled directly by some Java applets and browser plug-ins such as MapGuide, spatial queries are handled in a consistent manner. The coordinates of the mouse clicks are sent to the server along with any other criteria (entered in a form as previously described) to be processed by the GIS or DGI software, similar to the manner described for the recenter, zoom, and marker placement functions.

One drawback of the standard HTML interactive graphic methods (<img ISMAP> and <input type=IMAGE>, as described at the beginning of the Web Interface section) is that only one click is allowed. The coordinates are sent immediately, not waiting for another click (or more), which would be required for defining a rectangle, line, or polygon.

One way of circumventing this problem is demonstrated in the ImageNet service from Core Software (*http://www.coresw.com*), which can accept a rectangle (defined by two opposite corners) as a search criterion, as shown in the following illustration. When the user clicks on the first corner, it is submitted to the server, which returns almost the exact same page and map. However, this map includes a crosshair showing the location of the submitted point, as seen in the upper right portion of the map in the illustration. When the user clicks on this second map, the coordinate is used as the opposite corner of the rectangle.

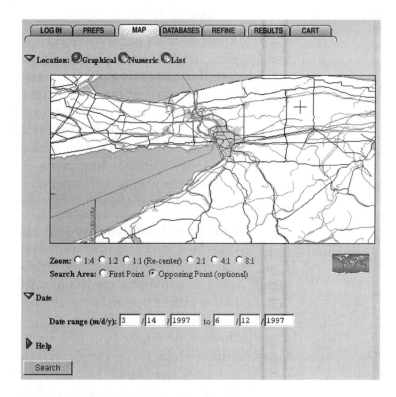

The ImageNet service, requesting entry of the opposite corner of a query rectangle.

Another solution is used by the NSDI Clearinghouse, which uses a Java applet for entering search criteria. Because the Java language is much more flexible than HTML, any necessary map interaction functionality can be incorporated into the applet. Thus, it is possible for users of the Clearinghouse to "drag" a rectangle on the map (i.e., push the mouse button at one corner, and hold it down as you move to the opposite corner, where you release the button), which they can submit as part of the query.

Advanced GIS Operations

Another level of functionality you may wish to add to your DGI interface and application includes more complex analysis functions, such as buffers, overlays, and reclassification. This type of feature will usually require the interface to be more flexible than the other tools described in this chapter. Although it is possible to create "canned" applications using these operations, it is more common to give users more general access, so that they may use analysis tools to explore your information for their own applications.

Enabling this flexibility will almost always make the interface much more complex. Each command will have a different set of parameters, for which you will need to provide a means of entering them. For example, a buffer operation will need to know at least the features to be buffered (e.g., selected from a list of themes, or pointed at in the map), and the distance to buffer. However, a reclassification operation will need not only the theme but a possibly complex algorithm for determining new classes from existing attributes.

This variety cannot be created using normal HTML forms, which are static (i.e., you could not have the bottom half of the form change on the fly according to what the user entered in another part of the form). The GRASSLinks service previously shown accomplishes this flexibility by using multiple pages. When the user selects one of the analysis operations, they are transferred to a separate HTML page, which includes a form for entering the required parameters. For example, the buffer page is shown in the following illustration.

Entering buffer parameters in GRASSLinks.

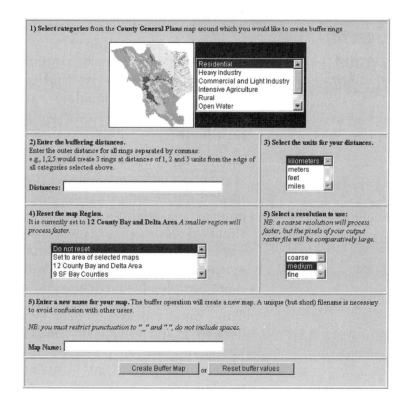

Another possible method for implementing a complex or conditional interface is to use a Java applet. Again, because the language is flexible, you can add whatever interface functionality you need. For example, the applet delivered with the Caris Internet Server (described in Chapter 4) provides for some analysis capabilities, and can be extended to include other functions.

Summary: General Design Guidelines

There are several general factors to consider when creating the interface to your DGI applications. The appearance, action, and relative location of each of the elements of your interface must be well determined in order to create a service that is both easy to use and powerful, and thus popular.

How Much?

The first general consideration is the overall amount of functionality to include on the page. Including more navigation, configuration, and analysis options increases the flexibility and power of the interface, in that experienced users have many ways of accomplishing exactly what they wish. This is the approach taken by the TMS. However, a full-featured interface can be more complex for novice users to learn.

Thus, mass market services such as online telephone books (e.g., the Ameritech Yellow Pages, shown in the following illustration) often opt for very simple interfaces. These include only the most intuitive and commonly used tools, designed to be very easy to understand.

Ameritech Yellow Pages has a very simple interface to accommodate non-geographic users.

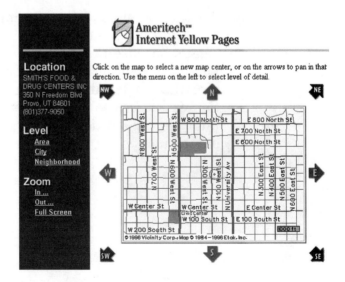

A good rule of thumb is that a service for casual users should be able to be displayed on one screen, without the user having to scroll (except to see auxiliary information such as help and links to other pages on your Web site). More complex interfaces can be up to two to three screens long because advanced users are more motivated to scroll around your page.

Tool Appearance

You must also design the appearance of each element to use for each function in your application. As previously stated, the main types of design elements that can be used include map images, graphical icons and buttons, text with or without hyperlinks, and forms with various types of input fields. Some of the common uses of these types of elements are described in this chapter, but there are some general rules you can follow.

Buttons and other graphical icons can greatly improve the appearance of your site, and can be very intuitive (e.g., a right arrow for "move east"). The appearance of the graphic is absolutely vital. If the icons are not well designed, or attempt to represent very complex concepts (e.g., "reclassify"), users will only be confused by the pretty interface.

Depending on the function, a simple text link is almost always more clear than a graphic image. For example, the words "overlay two themes" may be much clearer than trying to represent this concept in a small graphic image. When writing the text elements of your interface, it is important to use words that will be understandable to your intended audience. Technical, platform-specific terms such as "coverage," "design file," and "arc" may be perfectly clear to GIS professionals, but have no meaning for those not experienced with a particular brand of GIS.

Another important use of text in your interface is to explain the service and the interface to your users. This must also be carefully written so as to give them the information they need as concisely as possible (Web users tend not to read large blocks of text if they are in an otherwise visual page).

The most important uses of form elements are in cases where users must send information back to the server (e.g., the text of a marker label), or where they must enter several parameters before sending them all back simultaneously. Form elements can include not only text entry but lists from which users can choose one or more options. They might also include checkboxes.

Arrangement of Interface Elements

When designing your interface, the location of tools relative to the map is important, as shown in the following illustration. The first step in arranging the interface is to determine the element that should be the focal point (i.e., what users should see first and use most often), and place it in the most prominent position of the HTML document, which is usually the center of the first screen (more specifically, just above and to the left of center). The focus object may be a list of records, or a query form, but in most DGI interfaces it will be a map.

A generalized interface for a map browser and analysis service. Shading indicates relative importance.

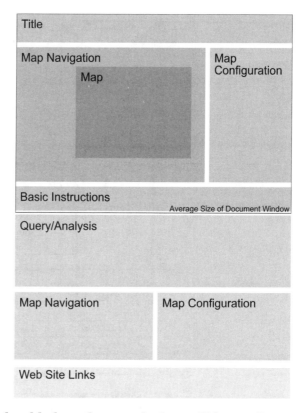

You should place those tools that will be used most often or are most important, especially for novice users, near the focal object. Because browsing the map involves a more direct interaction with the map image than the other types of tools described in the material that follows, the map navigation tools will usually be placed closest to the map image, either surrounding it (e.g., the Yahoo! Yellow Pages, *http://www.yahoo.com*, shown in the following illustration) or to one side.

Map navigation tools surrounding a map in the Yahoo! Yellow Pages.

Raxco Inc & Axent Technologies, 371 E 800 S, Provo, UT 84606(801-224-5306)

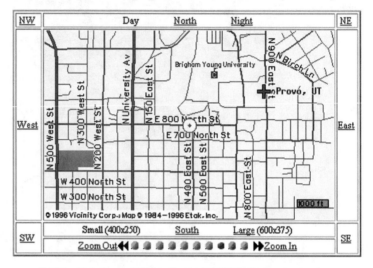

[BACK to Main Categories] | Print Preview | [SEARCH from another location]

[Questions, comments and suggestions. | FAQs | Need Help?]

Map configuration tools should also be near the map, but not as close as navigation tools. If there is room to one side, it can be used, or these elements can be placed below the map. Usually, analysis and query tools can safely be placed toward the bottom of the interface, where users can find them if necessary. These might also be placed on a separate page.

Designing a DGI interface, especially in the case of an analysis-capable service (which might consist of several pages), can require a considerable amount of work. Fortunately, many of the commercial programs described in Chapter 4 include extensive pre-built interface tools, which provide analysis functions. These can be customized to fit your application.

7

Get to Work

Implementing Your Site

At this point, you should have a detailed plan and design for your DGI services. Now it is time to implement that plan. As discussed in Chapter 5, it is likely that the implementation phase will take more time and money than the rest of your project combined. This chapter is intended to prepare you for this process, and step you through it in such a way as to save as many resources as possible.

Four phases of implementation will be discussed here. First, you may need to prepare your data so that it can be delivered by your DGI software. The next step is the development of your DGI application, including installing the DGI software and configuring and/or programming your interface and the engine that runs it.

Then there are other pages and documents that should be included, such as documentation for using the service and metadata to describe your information to users. Once your site is finished, there are several things you should do to attract your potential audience to your new site. These are also discussed in this chapter.

This chapter also includes guidelines for the long-term maintenance of your site, which is vital if you plan on your site being permanent. However, maintenance is often neglected until a crisis arises that could have been avoided or minimized with proper maintenance.

Data Preparation

The first step is to make sure your data will be available to the users of your service. In some cases, this will require very little effort, whereas other situations require continual effort to put your existing data online.

File Location

In almost every situation, the physical location of files is not very important. Using common LAN file sharing (e.g., NetWare or Windows Networking), it is not difficult for your Web server and the DGI software to see your data in its usual location (i.e., where the normal GIS applications access it). This has the advantage that the Web service will always be using the most up-to-date data.

However, this is not an ideal solution. In most cases, you will only want Web users to be able to access a portion of your information, not the entire data store. For most of the software packages discussed in Chapter 4, you have the ability to pick exactly which themes to make available. The rest remain invisible. Some of the programs even allow you to control various aspects of access for each user and each data theme, such as whether they can view, alter, analyze, or query the data.

However, there is still a small potential security hazard. Although you should be able to do this safely, any security issues should be addressed. The alternative is to have a separate data storage area to be accessed by the DGI software.

If you are using a firewall, it is likely that both the Web/DGI server and this redundant data will be outside the firewall, leaving the entire inside of the organization untouched. If a hacker were to damage this data, you would simply recopy it from the main data server. The main disadvantage of this approach, however, is that the entire data set must be copied to the public data store periodically to keep it up to date. Between these dumps, the information available to Web users will be to some degree out of date.

Data Format

One of the major advantages of the DGI programs produced by the major GIS software vendors (ESRI, Intergraph, MapInfo, and Genasys) is that they are able to access your GIS data in its native format (although in most cases you must buy the DGI program from the same vendor as your GIS software). That is, your DGI application can access the themes directly without any conversion. Possibly because of this convenience, and the general pricing strategies of the GIS vendors, most of these are by far the most expensive solutions.

Unfortunately, most of these native formats are proprietary, and specifications have not been released to the public (Intergraph/Bentley's DGN format being one exception). This means that third-party programs, such as Fornet Map Server and FME, can only access data stored in a public format. Although they may include many formats, including the GIS vendors' own exchange formats (e.g., ESRI's generate and shapefiles, Autodesk's DXF, and MapInfo's MIF), you must still use your GIS software to export the data into one of the supported formats so that it can be read. Thus, you are essentially creating a redundant data store, as discussed in the previous section, with all of its advantages and disadvantages.

In fact, some of the DGI programs, such as MapGuide, have their own proprietary format, and you must import the data from one of the exchange formats; that is, there is a two-step conversion process. This can take considerable time and effort, unless you create scripts to perform this automatically from time to time. Another potential drawback is analogous to the translation of literature between human languages: when data is converted from one format to another, some information is frequently lost, such as attributes, special object types, or positional accuracy.

Projection/Coordinate System

The last form of modification you may have to perform is to alter the map projection, coordinate system, or even horizontal datum. It is vital that all of the themes in the map be displayed in the same projection, so that the features will be aligned with one another in their proper relationships.

In most cases, all of your data will be in the same standard coordinate system, such as Universal Transverse Mercator (UTM) or State Plane (SPCS). Therefore, the relationships among objects are not a problem. However, if you are combining data sets from various sources, you may need to re-project some so that they all match, or use one of the few commercial solutions that can perform projection changes on the fly (such as Intergraph's GeoMedia).

The best long-term solution my be to store everything in geographic coordinates (i.e., latitude/longitude)— for many reasons. First, it is the only standard coordinate system that seamlessly covers the entire globe, and can thus handle any data set. Second, it is easier to do projections on the fly for display on the screen and for performing measurements. Also, the choice of projection and parameters can be optimized for exactly the area currently displayed—rather than for the entire data set—increasing the positional accuracy of the map. Several DGI programs can do this, including MapGuide (which actually requires data to be in geographic coordinates) and ArcView Internet Map Server. The geographic coordinate approach is probably more apropos for map browsers (where the area being covered by a single map will vary widely) than GIS analysis services, in which the map extent tends to remain constant and maps tend to be similar to traditional GIS applications.

Software Development

This phase of development is probably the most difficult, and most time consuming. The situation of each reader is very different: different goals and strategies, different applications, different choice of hardware and DGI software, and so on.

However, there are several general guidelines you should follow as you develop your DGI applications. This section discusses the common pieces of a DGI program, and the general procedure to follow to create or install one.

Creating Your Own Software

Most readers will probably decide to use one of the commercial or public domain DGI packages described in Chapter 4 (or perhaps one not covered) as the nucleus of a site. However, as described in Chapter 5, it is possible for you to create your own application from scratch. Although it will certainly require more work and some programming expertise, you could create a program more closely tailored to your application than many of the commercial platforms. Because only the necessary pieces of code are included, it may even run faster.

Even some of the commercial solutions, such as ESRI's MapObjects Internet Map Server, are only toolkits, not finished products. You still need to create your own DGI program, but the toolkit provides many of the pieces for you, which can save considerable time and effort.

As described in Chapter 4, there are two general types of DGI programs used for most applications. The map generator—a block diagram of which is shown in the first of the following illustrations—works directly with the data, performing all analysis and mapping itself. The GIS gateway (a block diagram of which is shown in the second of the following illustrations), on the other hand, is merely a live link between the Web server and a running copy of your normal GIS software. Map generators require you to create all of the GIS functionality yourself, but they are usually more efficient and run faster than the alternative.

Block diagram
of a map generator.

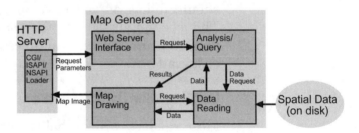

*Block diagram
of a GIS gateway.*

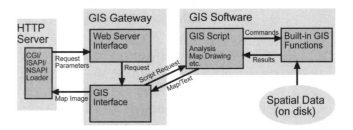

As you create your map generator or gateway, there are several elements that may need to be included in the program. These are described in the sections that follow.

Web Server Interface

The user's Web browser will not communicate directly with your program, but with an HTTP server. Thus, the DGI software must be able to communicate with the HTTP server, by accepting parameters sent with the request and sending the results back (whereupon the HTTP server will send them directly back to the browser).

The simplest form of interface is the CGI, which is supported by most Web servers. The request parameters are received either as command-line parameters or through standard input, and all results (graphics, text, and so on) are sent to standard output.

In the case of the NSAPI and ISAPI interfaces, your program is actually a dynamic library, which is loaded by the server when the request comes in. Your library consists of one or more functions that take the request parameters as input and return the resulting graphics or text. This method is usually much faster than CGI, but is not supported by every brand of Web server.

Most of the toolkit-type commercial solutions described in Chapter 4 include a ready-made function for this module, which you include in your program. For example, the Internet Map Server for ESRI MapObjects consists only of this module (the rest of the functionality is in the rest of MapObjects), and includes two pieces. One piece connects to the Web server via ISAPI or NSAPI, whereas the other is built into your program. The two communicate via TCP/IP, allowing for a distributed GIS server approach, as described in Chapter 5.

GIS Interface

There are several ways to communicate with your GIS software, with varying degrees of efficiency. The most simple method is to have the gateway program issue the commands to start the GIS program and run a script to do the specific processing; however, this process can take five to ten seconds or more, and is not recommended.

A more streamlined process is to use a standard form of live, inter-process communication, such as Dynamic Data Exchange (DDE), Component Object Model (COM), or UNIX Inter Process Communication (IPC). Here the GIS software is constantly running, waiting for requests. When your gateway program receives a request with parameters, it sends a message to the GIS program asking it to start a script to process the request.

This is the method used by many of the commercial live GIS interface platforms described in Chapter 4, and is relatively efficient. One drawback is that the GIS program, because it is constantly running, uses a considerable amount of RAM and other resources. Another drawback is that the solution is not scalable, because most GIS software can only process one request at a time.

A more robust solution, used by MapInfo ProServer and other programs, is to use a network-capable communication scheme such as Distributed Component Object Model (DCOM), Common Object Request Broker Architecture (CORBA), or Remote Procedure Call (RPC). These are similar to the previous solution, except that messages can be passed to GIS programs on one or more separate computers on the network. To handle more simultaneous requests, more GIS server computers can be added to the cluster.

As the script receives results from the GIS, it passes them on to the HTTP server, which sends them to the browser for display. It may do some further processing on this information (e.g., converting a text report into an HTML page), but this is not always necessary.

GIS Script

If you are creating a GIS gateway service (as opposed to a map generator), you need a way to instruct the GIS software on exactly how to process the request. This is normally done with a script, written in the native scripting language of the software, such as Arc Macro Language (AML), MicroStation Development Language (MDL), Avenue, GenaShell, and MapBasic. These are high-level languages that are usually easier to learn and use than standard software development languages such as C.

When a request is sent from the gateway to the GIS program, it starts this script, which has been created specifically for this application (although it may be very similar to applications you normally run on the GIS). The role of the script is to use the parameters to perform one or more GIS operations, generating some type of output to send back to the gateway (and eventually back to the user's browser), such as a map or a report.

The analysis/processing part of the script functions just like any other you might create to automate common tasks in your GIS, such as highlighting features on the screen that match a search criteria. The only difference is that the input is taken from the transferred request parameters (similar to when the script is called from the command line) rather than from a dialog box or mouse click on a displayed map.

The other major function of this element is to generate some type of output. Whereas the standard GIS script sends its output to the local screen display, or generates a new data theme from the resulting features, the DGI script must direct output to a very remote display, the user's Web browser.

This is done in one of two ways. Either the map graphic or text report is transferred directly back to the gateway program that originally called it, or output is saved to a file that is then read by the gateway and transferred to the user. The former solution is faster, but only works with some of the direct communication interfaces described above under the heading "GIS Interface."

The latter solution is not only slower, but can be a file management headache. First, the file names of the output for each request must be unique, or simultaneous processes might conflict with one another. Unique file names are usually created from the process ID or the user's IP address. This causes its own problems because the output files generated by each request will start to accumulate on the server very quickly. It is common to employ a small, separate housecleaning program that automatically removes all of these files every few days.

This idea of generating files can actually be a benefit in some applications. For example, the GRASSLinks service saves the output theme (not just the map to be displayed) from each analysis, as shown in the following illustration. These saved themes can then be used in subsequent analysis requests, which is invaluable in a process-type task that requires several separate operations in sequence. For example, you can buffer a certain type of wetlands, then combine the resulting buffer theme with a zoning theme to create a new theme that classifies the danger of encroachment. These output themes can also be downloaded by the user so that they can be used in the user's GIS software.

Saving the output of an analysis in GRASSLinks.

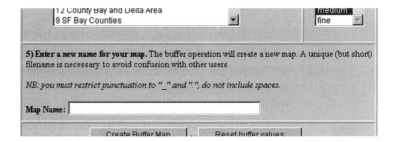

Reading Spatial and Attribute Data

In order to do any spatial processing, you must be able to access your spatial data. In a GIS gateway, your script (previously described under "GIS Script") is able to use the relatively high-level data reading functions built into the GIS software. However, if you are writing your own DGI module, you will need to read the data yourself.

Thus, you will need to know every detail of the format in which your data is saved on the disks. Many format specifications—such as Bentley's DGN, ESRI's Shapefiles, and Autodesk's DXF—have been publicly published, and some even have free libraries you can incorporate into your system for reading the specifications. On the other hand, some formats—such as

ESRI's ARC/INFO Coverage—are not public, and data can only be read by figuring out the format yourself (or purchasing an access library from the vendor).

Some data formats tend to be more efficient than others for live map generation, especially if your data set is very large. For example, the TMS does not read data from the standard TIGER/Line format, because it is very inefficient for real-time mapping. Instead, a format developed by the Environmental Protection Agency was used, which is considerably faster due to its binary format and spatial indexing.

Your application may also need to access attribute data (or even spatial objects) stored in a relational database. To search and read this database, you will either use the Application Programming Interface (API) from the database vendor, or a standard interoperable database interface, such as Open Database Connectivity (ODBC) for Windows NT, which provides a single interface to many brands of database systems.

Spatial Analysis and Query

This module is where the real work is done. Many applications will need to perform searches, classifications, and even spatial analysis operations according to the user's request. One exception is the dynamic map browser, which often offers no analysis capabilities and reads data directly to the output module. However, even some of these offer analysis functions such as finding optimal routes.

This analysis module can include almost any functionality you need, as long as you are willing to program the operations yourself. One of the advantages of using a GIS gateway approach rather than a standalone program is that in the former the GIS script can use the commands and tools built into the GIS, rather than having to recreate this functionality.

Another possibility is to purchase a commercial GIS library to give you this functionality. This includes the toolkits discussed at the end of Chapter 4, including ESRI's MapObjects (with or without the Internet Map Server module), Geodyssey Hipparchus, and others.

HTML Generation

In most cases, a DGI service consists of not only a map image but an HTML page containing other interface elements, as described in Chapter 6. This page is usually not a static file saved on the disk but is generated dynamically by a CGI script, based on many of the same request parameters as the map itself. In fact, when users access your service, they are actually using the URL of the script that generates this page, which is usually separate from the map generator/GIS interface script discussed in the rest of this section. This script separation is shown in the following illustration.

HTML generator and map generator scripts.

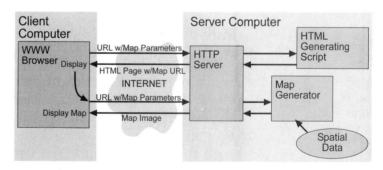

Essentially, the script creates HTML code, which is passed back to the browser as if it were a normal file. There will probably include constant elements (e.g., title, corporate logos, and links to other services on your Web site) that will be printed without modification. However, much of the code will be created using the parameters.

For example, if the interface includes a left arrow to move the map to the west, the graphic will be constant, but the <a> tag that links it to the new map page will be different every time, depending on the parameters the browser sent with the request for the current page. That is, the center longitude parameter of the link is slightly less than the center longitude that was passed to the script.

If a dynamic map is included, the <img> tag will also be generated dynamically. The basic URL address of the map generator program will be constant, but the same parameters (the HTML generator script) that were passed to this will be passed on to the map generator as URL parameters.

An alternative to a CGI script is to use an inline macro language, which allows program code to be included in an HTML file, and processed by either the server or the browser whenever the page is requested, modifying the resulting HTML code. Netscape's JavaScript MicroSoft's Active Server Pages, and Internet Factory's SMX are popular and powerful examples.

It is also possible for this module to be one part of the overall analysis or GIS interface program, provided the output is text rather than a map. For example, a user might perform a search for geographic features, with the results being a list from the database. This module would take a plain text report and add HTML tags (e.g., for headings, bulleted lists, boldface type, and italics) to make it more readable.

Map Drawing

The final module of any map-based DGI program is the part that creates the map image to send back to the user. If you are creating a GIS interface service, this is handled by the GIS script, which uses the normal plotting tools of the GIS software to create a map (usually saving it to an image file as described in the GIS script section).

If you are writing a map generator, you will need to create this image by drawing to an image structure (e.g., in GIF format) in RAM memory, sent back to the Web server through standard output for CGI or, in the case of NSAPI or ISAPI, returned from the main function.

Creating this image can be rather difficult if you have to write the drawing code yourself. Fortunately, public domain libraries are available for most common formats (such as GIF, PNG, CGM, and JPEG), which you can incorporate. Thus, you would issue a command such as *drawline(10,10,56,45)*, and the drawline function from the library would worry about the formatting details.

Using a Ready-made Product

The alternative to doing all the work yourself is to use one of the commercial products described in Chapter 4. These products are similar to the applications you could build yourself (as described in the previous section), in that they include some or all of the modules previously mentioned. The difference is that the vendor has already combined the modules into a complete program (or set of programs). They have made efforts to automate as much of the service as possible, including Web and GIS interfacing, data reading, and map drawing, relieving you of these responsibilities.

Although these pre-built solutions can save you a considerable amount of work, none are truly turnkey solutions. All of them require you to at least configure the engine and the interface for your particular application (as developed in chapters 2, 3, and 5), and the design created in Chapter 6.

This section describes several steps you may have to perform to build a working application using a pre-packaged DGI platform. However, the detailed process will be drastically different for every brand of software; therefore, you should consult the manual for your software for technical details.

Software Installation

First, you will need to install the software on the proper machine. At least the majority of it will be placed on the Web server, whereas parts may need to be installed on other computers if you are using one of the three distributed approaches described in Chapter 5.

The first part of the process is to physically place the software on the computer's disk. For most of the commercial platforms, this process is automated, and relatively simple. However, some of the packages, especially the public domain UNIX offerings such as ForNet Map Server and GRASSLinks, are distributed as source code that must first be compiled. This can be inconvenient and sometimes difficult, but it does give you the opportunity to alter the software if necessary to meet your needs.

The second part of this step is to configure the interface between the DGI software and the HTTP server (this interface is discussed in the previous section). Even ready-made software must include this configuration. This can be very different for each combination of these two elements. If a CGI interface is being used, very little work is needed: in the Web server configuration, you need to make sure that the server recognizes the DGI program as a CGI script so that it will be executed rather than downloaded (as are most files on the Web server). This is usually done by placing the program in the *cgi-bin* directory with the rest of the CGI scripts, but it is also possible to tell the server how to access the program, no matter where it is.

For ISAPI and NSAPI interfaces, the process is somewhat similar. The Web server needs to know where the DGI program is, and that it is to be executed with the appropriate method (ISAPI or NSAPI). You may also need to specify which types of requests should be handled by the DGI program.

The last part of this step is to configure the GIS interface portion, described in the previous section. Most importantly, the gateway program needs to know how to communicate with the GIS software, via a live link, or how to start the program when needed.

Once this step is complete, you should be able to run the DGI program from a Web browser by requesting the appropriate URL. It will not do anything useful yet, but it should at least respond if everything is working. Many of the software packages (e.g., Autodesk's MapGuide) described in Chapter 4 have test scripts or parameters that aid in this check.

Application Configuration

The next step is to set up the software for your particular application. The package as it ships includes the skeleton of a DGI application, including the communication protocols, the analysis and mapping functions, and the ties to link everything together. However, you must add the details of your design, including the appearance of the interface and the exact tools to be made available to users.

For some of the programs, especially the map generators, this is relatively simple, involving only a few configuration settings. The variables you need to set
include the following:

❑ The location of the spatial data themes and tabular databases to be available (either a directory on
   your local disk, or a remote location if the data is
   distributed).

❑ The visual tools to be included in the interface
   and their locations and actions (as discussed in
   Chapter 6).

❑ The design of the output maps, including the symbology of each element, and the scales at which
   each element should be visible.

❑ The names and locations of scripts that handle
   more complex tasks, such as spatial analysis (see
   the following section).

Setting variables is done by either editing a configuration file or using a GUI program with dialog boxes and
other interactive elements. An example of the former
is the ForNet Map Server: a text file contains all of the
basic settings, whereas the appearance of the interface
is controlled by template files. The template files are
similar to HTML pages, but with portions left blank to
be filled in by the dynamic map generator.

Autodesk MapGuide is a good example of the latter.
The generator software (the MapGuide Server and
MapGuide Agent programs) has a GUI interface that
includes a capability for setting the necessary parameters, and tools for managing and tracking the activity
of the service as it is running. MapGuide Author is the
third piece of the package. It is a GUI application for
designing the maps to be sent to the user. MapGuide

Author is very similar to a desktop mapping program (e.g., MapInfo or ArcView) in appearance and usage, making maps relatively simple to design.

In order to make setup as easy as possible, these programs usually sacrifice flexibility and functionality. That is, when there are fewer tools available, it is invariably easier to use them.

Application-specific Scripts

In order to include more functionality in your application, such as GIS analysis, you will need to provide more details on the tools you wish to include and how they perform their tasks. This involves creating two types of scripts (i.e., high-level programs).

The first is the GIS script, exactly the same as in the previous section. This script uses the functionality of the GIS software to process specific requests. (Normally, different operations available in the Web interface would access separate scripts.)

The advantage here over the do-it-yourself scenario previously described is that most of the commercial products provide many aids to you in creating these scripts or adapting existing GIS applications for this medium. They may have sample scripts you can adapt to your situation, or extensive documentation, or even extensions to the GIS software that automate the communication and map generation tasks. This is how the ArcView Internet Map Server is used.

The second type of script is at the other end. Some products, such as GeoMedia Web Map, consist of the analysis and map generation modules (essentially a small GIS program optimized specifically toward this type of application), but without the Web server interface portions. Thus, you must create your own CGI scripts, which generate the HTML interface (as described in the previous section), receive the requests sent by the browser, and send those requests to the map generator/GIS analysis engine.

When maps need to be included directly, it may be possible to access the analysis/mapping program directly. This segmented approach is shown in the following illustration.

Architecture of a service in which the analysis/mapping engine and Web interface script are detached.

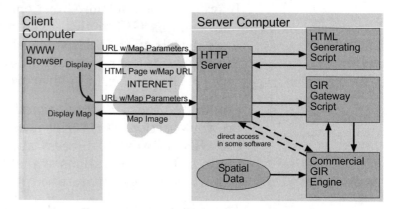

As with the GIS script, the purchased package usually includes some aids for creating the necessary CGI script. This includes sample scripts and documentation on how to access the engine. In fact, some vendors (such as Intergraph and Object/FX) include consulting services in the package, in which their personnel build the necessary scripts for you, for a fee.

Once these first two steps (data preparation and software development) are complete, your site should be usable. At this point, you should do extensive testing of the services to be sure they work and have no bugs.

Testing the service consists of trying out each of the tools in the interface to see if they operate as expected.

It is good to test them in multiple situations (e.g., when the map is in a different region, at different zoom levels, or when different themes are being shown). Each of the data themes should also be checked to see if they appear correct. In addition to technical testing, you should also have some potential users try the service, to get opinions on the quality of the appearance and the usability of the service.

It is very common in the Internet community to release alpha- and beta-level services before testing has been completed (or even begun). This may allow you to get a jump on possible competitors. Experienced users who are accustomed to this practice will understand its status and be helpful in reporting bugs and making suggestions.

However, as the Internet community grows, most people not accustomed to the unique practices of this environment have the justified expectation that services and software that have been released should work without problems. If there are bugs in your service, or if it does not operate as expected, these people are not likely to return to your site. Also, if your service is part of an intranet, it may be a mission-critical application, in which important activities will depend on the software working without problems.

Auxiliary Documents

Dynamic maps and analysis tools are not the only pieces of a complete DGI service. There are many other files that although less glamorous are nonetheless very important. Three types of textual documentation that should be integral parts of your service are an interface help, metadata, and a subject encyclopedia. These are described in the following sections.

Interface Help

No matter how simple you make your interface, there will be some users who will have difficulty using it. This is because everyone approaches your service with different expectations and mindsets. As with many other forms of organization (e.g, a filing cabinet), what makes perfect sense to you (e.g., what an icon represents) may be very different from another person's—even another expert's—understanding of the same concept.

Thus, it is important that you provide good instructions that describe how to use your service. Although it is possible to include complete details within the interface itself (e.g., next to the Zoom In button, having a sentence describing its purpose), it is generally not a good idea to sacrifice the conciseness of the interface. If you do, it will be more cumbersome for experienced visitors to use, and Internet users have a general tendency to avoid pages with a lot of text unless they really need the information.

It is therefore best to have extensive help as a separate document (or set of documents), with links from the interface pages. This linking can be done in several ways, such as a single link at the top of the screen ("for help on using this service, click here"). Another option is to include a help link at each interface element that connects directly to the section describing that element (marked with the <a name> HTML tag). The link could be a word such as *Help* or the standard question mark icon. The documentation should include at least two types of information for each part of the service.

❐ Explain the concepts behind the element; that is, what its purpose is. For example, "ZOOM IN increases the scale of the map; features appear larger, and more detail is included, but the map covers a smaller area."

❐ It should explain the manner in which the tool should be used. For example, "To zoom in, select the checkbox next to ZOOM IN, type in the desired zoom factor in the box to the right, then click on the map at the location you want to be the center of the new map."

It is also very useful to include one or two complete sample applications, especially if the service is complex. That is, include complete instructions for an example that will generate results. For example, "Click on the buffer tool; for the buffer theme, select wetlands, and enter '100' for the buffer distance (in meters); submit, and the wetlands and resulting buffer will be displayed in the map."

This sample has several positive effects. First, seeing real results will encourage your users, who may get quickly frustrated if they cannot get results rapidly. Second, it will help them see what they should expect for results when they try their own parameters. Third, it is generally much easier to modify your example parameters to meet their needs than it is to come up with correct parameters from scratch.

Metadata

The second piece of documentation is a description of the data, called metadata. A metadata record lists several characteristics of a data theme (e.g., the streets layer). This information includes the following.

❐ **Footprint/Spatial Coverage**. The geographic
area covered by the data set, defined by either a
rectangle (easternmost and westernmost longitude, northernmost and southernmost latitude) or
a more complex shape, as shown in the following
illustration. It is also possible to use recognizable
place keywords for such things as administrative
units or important features included. The projection and coordinate system of the data set also
falls under this category, as does the resolution of
raster data sets (the geographic area covered by
each cell). This is probably the most commonly
used element of metadata because it is a central
criterion for potential users.

*Means of
describing
the footprint
of a data set.*

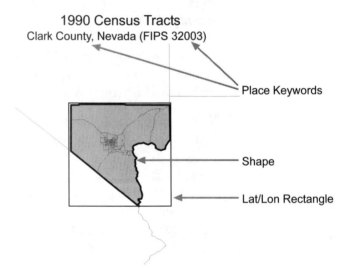

❐ **Thematic Coverage**. The geographic subject matter represented by the data set. The detail may
range from simple keywords (e.g., streets, geology,
or rivers) to more complete textual descriptions. An
important item is the level of detail included (i.e., a
transportation data set may include every street,

or just major highways). Some description of available attributes is also useful, as is an indication of the time period represented.

☐ **Data Format**. A description of the formal structure of the data. It may include a choice of standard format (e.g., SDTS, DXF, or ARC/INFO EXPORT), or a precise specification for the format. This portion should also include simple statistics on the contents, such as the number of objects for a vector structure, or the number of rows and columns for a raster structure.

☐ **Lineage**. The source of the information represented in the data set. This may include such items as a source document bibliography, field notes, description of the entry process, and copyright notices.

☐ **Data Quality**. Perfect accuracy is impossible in any representation of reality. Therefore, some indication is important of what level of positional and attribute accuracy can be expected. This is often difficult to measure precisely, but a rough estimate will usually suffice.

☐ **Contacts**. Information about the people and organizations responsible for the data set. This may include names, postal addresses, telephone numbers, e-mail addresses, and Web page addresses for creators, maintainers, public relations contacts, and salespeople.

☐ **Distribution**. Instructions on how the data set can be obtained, if this is permitted. In the context of this book, the Internet is at least one means of distributing the data, although a small raster map is not the same as the entire GIS database. This may include contact information for salespeople, an online ordering URL, prices, and restrictions.

Most government agencies in the United States and many other countries are required to create metadata for any data sets they create, but there are good reasons for any organization to do so, however much of a hassle it might be.

Most importantly, serious users can review the metadata before they begin using a complex DGI service, or before downloading or purchasing your data. Users can get a good indication of whether the data sets will work for their needs, possibly saving them a considerable amount of time and money pursuing useless data sources and possibly further encouraging them to use your data instead of another source.

Metadata also helps your clients make the best use of your data once they have obtained it. There will be some applications for which your information is especially well suited, and others for which it should never be expected to be useful.

Because of the wide variety of uses of metadata, creating metadata records for each of your data sets can be a daunting task. To assist in this process and to ensure that metadata will be intelligible by more people than its creator, standards have been developed (and are continuing to be developed) for the structure and content of metadata documents.

The most important of these standards to date is the Content Standard for Digital Geospatial Metadata (CSDGM) from the U.S. Federal Geographic Data Committee (FGDC). It is an extensive specification for exactly what elements should or may be included in a metadata record, and can be obtained from the FGDC Web site at *http://www.fgdc.com*.

Because CSDGM is so extensive (over 200 fields), it can be quite laborious to create records from it. Therefore, FGDC and other groups have produced many tools that help in creating records (which can also be obtained at the FGDC Web site). One very useful group of tools are scripts (for most GIS software) that generate many of the metadata elements (e.g., spatial location, attributes, and projection) automatically from the data set.

For large data sets distributed or sold primarily to GIS users, metadata should be relatively complete, including all applicable CSDGM fields. On the other hand, a mass market application still needs some metadata for interested users, but not nearly the detail. For example, metadata for a street map browser need only include basic information about the area covered (e.g., "United States"), source of the data (e.g., "TIGER 94 with our own additions from field work"), timeliness (e.g., "last updated April 1997 from field work September to December 1996"), and a copyright notice and use restrictions.

Once these records are created, you can link to them from the HTML pages of your DGI interface. You may include one link at the top of the page (e.g., "Detailed information about these maps"), or links from each theme in the legend or in the theme display control portion of the interface.

Subject Encyclopedia

The last form of documentation will not be necessary in many applications, but vital in some. In some cases, your information may cover very specific, complex concepts and issues, with which some of your users are not familiar. This is especially the case in scientific or technical applications, but may arise in many different situations.

Thus, this form of documentation provides a basic introduction to the subjects your information covers. In essence, you are teaching your users about what you are doing. In many cases, you may already have these documents if the DGI site is presenting information about the mission of your organization. In other situations, you may need to write these documents especially for the DGI service.

For example, an environmental activist association may have a map browser showing endangered species habitats being encroached. Although most people hear about these issues in the media, they probably know very little about the science or the laws involved. Education of this sort is a central mission of most activism groups.

Thus, you may include documents describing the species of concern, the legal and scientific meanings of such terms as *endangered* and *habitat*, and the specific encroachment risks (e.g., loss of water, loss of vegetation, separation of winter and summer ranges) faced by each species.

A paper company may use a similar set of documents as part of a DGI service to help tell its side of the issue. This will help your users make better use of your information, and help you make your point more effectively.

Obviously, if your audience is expected to consist primarily of experts in your field (as described in Chapter 2), or if the subject matter is simple enough (e.g., streets), this will probably not be necessary. An intranet application, where personnel should already be proficient in their tasks, should also be exempt from this type of documentation.

This encyclopedic information can be presented in many forms. You may have it broken down into very specific help items (i.e., "What does this mean?" links), or you may have a sequential "Guided Tour" of the subject that potential users can follow before they begin using your service.

Promotion

At this point, your service should be created, tested, and documented. It is now ready for your users. Your potential audience may range from five to several million, but regardless of its size, you need to let them know the service is ready and invite them to come and use it.

One of the drawbacks of the Web is the difficulty users have finding a particular site. Just because a page exists does not mean interested people can find it, even if they know it exists. This is because the Internet is extremely large (there are well over a million individual Web sites), but more importantly because it is not inherently organized. The location of a page (i.e., its URL) has nothing to do with the subject matter or any other characteristic of that page (except, perhaps, for its creator, in that the page will usually reside on the creator's Web server).

The Internet is more like a city (in which people's homes are located randomly, rather than by occupation or last name of the occupant) than a library (in which books are located according to subject). Thus, you cannot expect users to accidentally find your new service. You need to put some effort into advertising in those places your intended audience will visit. The following sections describe several promotional vehicles you might find useful.

Using Your Own Promotional Outlets

A relatively easy means of announcing your new service is to use the promotional channels your organization currently has in place. You should already have media and other outlets you use for announcements and advertisements. Three types of these are discussed below.

Personnel in Your Organization

Promoting your site is easiest if you have developed your Web application for use on an intranet. In most cases, you already know the people who will be using the service, and it is easy to tell them about it. Even large organizations have newsletters and e-mail lists that reach all personnel. You can also organize training sessions to teach personnel how to use the service. This is often very important because intranet applications tend to be more detailed than sites designed for the general public.

Even if your site is intended for users outside your organization, it is useful to let other personnel know about the site; they may be able to promote the site through their own channels. Word of mouth remains one of the most effective means of promoting sites on the Web, especially to specific groups of users.

Your Web Site

If you already have a general Web site for your organization, it is likely that many of the people who visit that site will want to use your DGI services. Thus, the first step in promoting your new service should be to make links to it from your home page or other appropriate pages. It is important that the links make it clear exactly what the new service is and why users should try it.

Traditional Media

Almost any organization has a body of existing promotional literature, including newsletters, brochures, and magazine and broadcast advertisements. Because these vehicles already have a circulation audience, they can be a highly effective means of announcing your new service on the Internet. This is more than the "www.xxx.com" at the end of a TV ad; you can use these outlets to describe the service and explain how useful it can be to readers.

Placing articles in print media can also be a great means of promotion, which can be done without cost to your organization. Although you will have little chance of getting into the major GIS magazines and journals, your service may be unique in its realm. Thus, you may find that the magazines, journals, and newsletters that cover your specialty are very interested in showing their readers what you are doing. For instance, if a municipal government has a map browser showing detailed geographic information for the city, the primary audience is likely members of the local community. A locally distributed newspaper or magazine is an ideal means of reaching this audience.

Links from Other Web Sites

With the previous three promotional outlets, you are essentially pointing to (advertising) yourself. This method will reach much of your existing audience, but will not effectively announce your services to a larger potential audience. To do this, you need to get other Web sites to point to you, preferably sites that are visited most often by your preferred audience. Three general types follow.

Global Search Engines and Indices

When most people want to find something on the Internet, the first place they go is to one of the major "Web catalogs." These sites attempt to list most, if not all, Web sites in the world, allowing you to search or browse them to find the particular information you need.

These services fall into two groups. The first are databases of Web pages with little or no selection—every page is included, no matter how unimportant. The record for each page has keywords extracted from the text of the page, which users can search. These "search engines" include the following:

❐ Lycos at *http://www.lycos.com*

❐ Alta Vista at *http://www.altavista.com*

❐ Web Crawler at *http://www.Webcrawler.com*

Services in the other category are more like catalogs, where sites are listed by various subject categories. To use them, you follow links through their hierarchy of categories (e.g., Science, Geography, GIS, and How-to Guides). Included links generally point to complete Web sites, not individual pages (i.e., the home page for *www.xxx.com*, rather than every HTML page on that server). These "Web indices" include the following (Lycos and Web Crawler have categorized indices, as well as comprehensive search engines):

❐ Excite at *http://www.excite.com*

❐ Yahoo! at *http://www.yahoo.com*

❐ Magellan at *http://www.mckinley.com*

Because these are some of the most heavily visited sites on the Web, it is advantageous for you to be listed in them, with the proper keywords or categories so that your potential audience will find your site as they look for the information they need. Although most of these catalogs include robots (programs that automatically browse the Web) that will add new sites, the chances of your site being found accidentally within a few months are slim. Fortunately, all of these sites give you the ability to add your site manually. Just look for an "add your URL" or similar link.

A new form of specialized index is the NSDI Clearinghouse, which is intended to permit indexing and search online metamedia entries that point to online data or mapping systems. Unlike all-purpose Web indices, Clearinghouse supports search for systems based on geographic coordinates, time period content, and many other fields. Targeted at geospatial data and services, registration of your data and server with the Clearinghouse should be a consideration in building access to your data by other GIS professionals and data users.

Subject Indices

In addition to these global databases, there are indices created for specific subject areas (e.g., wildlife management, city governments, or businesses in Denver). These are usually maintained independently by people and organizations interested in a particular subject. Because of this vested interest, these indices tend to be well organized, comprehensive, and updated frequently. Thus, sites offering these indices are frequented by people highly interested in the subject, which makes these sites valuable places to have your service listed.

You should therefore find two or three indices that most closely match the subject matter of your service (or even the general geographic location of your information). Because these sites tend to be independent, they can be difficult to find. Two good places to look for them are in Yahoo! (*http://www.yahoo.com*), which has a category "Indices" under many of its subject headings to contain these lists, and the WWW Virtual Library (*http://www.w3c.org/vl*), which is a large collection of independent indexes. Once you find a few good lists, you should be able to find a form or e-mail address to use to add your own service.

DGI Indices

One of the subjects covered by these focused indices is DGI itself. There are several sites that do nothing but list examples of services that integrate GIS and mapping on the Internet, regardless of their subject matter. The following are three of the more popular and comprehensive services.

❐ Web GIS: Toy or Tool? Resource List, by Bill Thoen, GISNet, at *http://www.gisnet.com/gis/notebook/Webgis.html*

❐ Database Query and Map Generation on the WWW: Online Resources and Examples, by K. A. Duda, Michigan State University, at *http://www.ssc.msu.edu/~geo/wwwgis.html*

❐ Maps and References, by the University of Iowa CGRER, at *http://www.cgrer.uiowa.edu/servers/servers_references.html*

The companion Web site to this book, at *http//www.kayenta.geog.byu.edu/gisonline,* also contains a rather comprehensive index of DGI services. Each of these indices has a form or e-mail address to which you can submit your new site.

Maintenance

Although you will have spent at this point considerable time and effort creating your Web services, your work is far from complete. Because you have invested so much, you want your new site to be used for a long time. Thus, you will need to continue to invest in its maintenance.

This phase will require considerably less work than the initial implementation, but there is still a lot to do. Maintaining your service will consist of several tasks that need to be performed periodically. These are described in the following sections.

Keeping the Service Running

Unfortunately, Web servers have not matured to the point that they will run indefinitely without crashing. Periodically, you will have problems with the server computers, the network, the HTTP server software, your GIS software, and the DGI applications.

To manage the range of potential problems, you will need to have permanent staff who are experienced in each of these areas and who can correct problems. If you have temporary people building the site, make sure that permanent people are assigned to work with them to learn the technical details of how the site works. These can be members of your computing personnel (for the hardware and HTTP server) or your GIS personnel (for the DGI application).

User Feedback

Although you will spend some time trying to identify your audience, their characteristics, and how your service should be structured to accommodate them, you will probably not get the mix and balance of these considerations perfect on the first try. It is inevitable that your users will have ideas (positive and negative) about your site, and how it could be improved. In many cases, the main goal of the service is to provide users with the information they need, so you should be responsive to their ideas.

It is advisable to include a link on the site that allows users to voice their concerns and make suggestions. This may be a form or an e-mail address. If using a form, it may be useful to include checkboxes to differentiate between different categories of comments: bugs and serious problems, general complaints, comments or additions to the data, suggestions for improving existing services, or potential new services. It is then possible to automatically forward each message to the appropriate member of your staff.

Of course, how you deal with these comments is your own prerogative. Some will offer great insight and help you improve the quality of your site; others will be offhand comments you can ignore. Either way, it can be very helpful (and perhaps a bit humbling) to hear what users think of your work.

Augmenting Your Services

Whether based on user feedback or on your own ideas and experiences, you will periodically want to improve or add to your DGI applications. This will require more work than just keeping the service alive, but not as much as the original implementation. There are several types of augmentation you may want or need to do.

❑ **Handling increased load**. As your site becomes more well known (and as the number of users on the Internet increases), your server may eventually become overloaded with requests. The bottleneck might be in the network connection. The problem might also be insufficient CPU power, RAM memory, or disk space, all of which can be upgraded for a price. If you have already upgraded your server to the limit and still have problems, you should consider the distributed processing approaches discussed in Chapter 5.

❑ **Upgrading hardware and software**. The inevitable fact of technology is that it is constantly becoming obsolete. Periodically, new versions of your computer hardware, operating system, HTTP server, GIS software, and DGI software will become available. These may correct problems with earlier versions, perform better, or add new features. In your situation, it may or may not be advantageous to upgrade, and you should consider whether it is worth the upgrade price.

❑ **Modifying the interface and functionality.** As you observe the usage of your site, you will notice that some features work well and are used often, whereas others are either not understood or not used by your audience. You should use these observations as well as user comments to make positive adjustments to the interface, the maps, and perhaps even the operation of your application.

❑ **Adding new features**. Whether originating in your organization, or from user feedback, there will be many ideas for new functions (and possibly entirely new applications) you will want to add to your site. The planning and design of these should follow the principles presented in this book, but will certainly not be as involved as the initial implementation.

Summary: Rolling with the Punches

This chapter has given you an indication of the tasks you need to perform to get a site up and running. However, how you use this information in your own implementation will vary depending on your situation and your plan. Some services will require extensive and continual data preparation; for others you will be able to use your existing data. Some applications and software will require much more work to create and install than others. Also, the extent of the documentation you will need to provide is highly dependent on the type of information you are distributing and your potential audience.

No matter how detailed your plan, things will change as you go through the steps described in this chapter. The software you buy may not work exactly as you expected. Some of the functions you would like to include may not be possible to implement. You may have new insights during implementation that improve on the plan. You should expect this, and keep your plan flexible enough to accommodate these and other eventualities.

The labor required will be the largest uncertainty. You will never be able to know exactly how much work is required in implementing the site until you do it. All of the possible solutions will require some work, and few are as easy as vendors advertise.

The process of creating and maintaining an online GIS or mapping application can be very tedious, frustrating, and expensive. However, if you have planned effectively, both for the design of your application and for the process by which you create it, it will be well worth it. You will likely save resources, prevent much of the potential frustration, and create useful and profitable services that can be maintained at a reasonable level of resource commitment.

8

The Future of DGI

Issues, Concerns, and Solutions

This book has discussed many ways you can build Web-based services that distribute geographic information to a widespread audience. As of the time of this writing, all of the solutions discussed can be implemented, and the applications you build today will likely work for a long time.

However, technology does not stand still, even in this field. It is very likely that new methods will become available in the near and distant future that will run your application faster and better, and allow you to include features not possible today. Technologies may arise that change the entire field, allowing you to distribute your information in ways totally undreamed of today. However, there are many ways in which the solutions you are able to build today are not ideal.

In addition to technological developments, there are other issues surrounding this subject that need to be considered and resolved before we can all smoothly and transparently share spatial information. This chapter looks at many of the shortcomings of the current DGI industry, and some of the solutions that are either being developed or could be developed.

Concerns with Technological Solutions

Although most of the issues discussed in this section are problematic, and require complex solutions, it is likely that many of them will be solved through improved technology.

In some cases, existing software may be improved or extended, whereas other solutions may require entirely new approaches. At least some of these potential solutions are currently under development, but some are only labels for needs that may one day be met.

It should be noted, however, that although a solution for each of these problems may be available soon, there is also an inertia effect to overcome. It takes time for software vendors, service maintainers, and users to all accept new technologies and fully incorporate them into their work, individually and collectively.

Performance

Although users are willing to wait for some of your functions to run (such as complex analyses), they will expect the more basic functions, such as map browsing, to be interactive, with almost immediate response. Ideally, users should not be able to tell the difference between a network-based application and a local application.

However, this is rarely the case, because most map generators take a few seconds to generate a map. There are several possible reasons for this, including the bandwidth of your Internet connection and the complexity of the software (i.e., how much memory, disk, and processor resources it requires).

Most of these problems will eventually be lessened as general computing and network technology improves, allowing for faster execution of programs and faster transfer of data. More efficiently written software and more compact data formats will also improve the responsiveness of your services.

Integration with Client-side GIS Software

One of the strengths of GIS is their ability to integrate many disparate data themes to find answers to complex questions. This data has to come from somewhere. If it can be obtained from an existing source, considerable time and money can be saved in data entry.

Thus, data download services (described in Chapter 3) have begun to appear, in which users can obtain existing data sets and use them in their own GIS software. Although useful for distributing data, this method is cumbersome because it requires many steps. Users must in turn find appropriate data sets, download them (in their entirety) over the network, convert them into a format understood by their software, and integrate them with the other themes they are using.

The data formats are especially problematic because each GIS platform has its own proprietary system for storing data. Sometimes you can find conversion programs to transfer data between these, but this is not always the case, and is inconvenient anyway.

Solution: Open GIS Formats/Protocols and Net-savvy GIS

The obvious way to get around problems associated with proprietary data is to have open data. If GIS software were able to read GIS data in a completely consistent way—regardless of the brand of either the data server or the GIS client, or where these are located on a local or global network—client users, server maintainers, and even software vendors would benefit greatly.

Data users could purchase one GIS program and access not only their own data for internal use but other sources, regardless of the software used. Data service organizations would be able to reach a much larger market than just the users who happen to be using the same brand of software they are. Both parties would be able to base purchasing decisions solely on the merits of the software itself, not on what the other side is using (a major factor in current purchases).

The Open GIS Specifications

There is a need for GIS vendors, who have control over both pieces of software, to cooperate with one another. This is being attempted in the Open GIS Consortium (OGC), in which all of the major vendors have both technical and management representatives. The OGC has developed an abstract specification to allow GIS data servers and processing clients to communicate with each other, whether on the same machine or across a local network or the Internet. This abstract does not define the precise protocols and such, it just specifies what a protocol needs to include.

Technical specifications, which implement the abstract requirements in each of the common distributed computing environments (DCOM, CORBA, Java, and the Internet), have been developed. In fact, they are already starting to appear in commercial software; namely, the net-savvy GIS programs described in Chapter 4. Intergraph GeoMedia uses Geographic

Data Objects (GDO), the DCOM solution it wrote jointly with many other vendors, whereas LAS GRASSLinks uses a TCP/IP (Internet) based solution it developed called Open Geospatial Database Interface (OGDI). When the specs are ratified by OGC, net-savvy GIS should rapidly become more prevalent.

Proprietary Map Viewers and Transfer Formats

Several of the vendors described in Chapter 4 use medium-weight (somewhere between a thick client and a thin client) client setups—based on a Java or ActiveX applet, or browser plug-in applets—that display maps generated by remote servers, and include other mapping or GIS interface tools. However, each of these applets reads maps drawn in different data formats and communicates with map servers using proprietary protocols.

These applets cannot read maps from servers from another vendor. The only exception is the ActiveCGM format (generated by both Intergraph and Bentley servers), which is publicly available.

Consequently, if Web users want to use several sites, they will need to download many viewers. Although this is not a problem for frequent users, casual users will not be willing to spend the time. This heterogeneity also makes it difficult for users to combine data from multiple sources.

Solution: Open Map Sharing Formats and Protocols

As with the problem of integrating GIS software-described above, the solution here should be standardization of viewers, so that people can use a single viewer to access any map source. This would be accomplished by transferring maps in a standard file format using a standard communication protocol, just as with the net-savvy GIS approach discussed under "Solution: Open GIS Formats/Protocols and Net-savvy GIS."

The difference between the two is in the form of information being transferred. In the former, raw spatial data—including geometry (whether raster or vector), topology, and attributes—is sent to the client software looking much the same as the data normally read by GIS software from a local disk. Processing, analysis, and map generation is done by the client.

In this map sharing approach, the processing is done by the server. The transferred file is basically a graphic image, with geometry, and sometimes simple attributes. The geometry is an extract of the original data covering the map area—perhaps generalized or otherwise altered to accommodate map display. Topology and other structures needed for analysis are not included, but map design information such as symbology, labeling, and legends is described in detail.

Standardization

The only current standards are the raster GIF and JPEG formats supported natively by Web browsers. However, these may not be ideal for many reasons, discussed elsewhere in this book. OGC has not in the past been involved in this process, but has recently begun looking at the issue of Web-based map sharing and how it may be solved technically. Unfortunately, the motivations for standardizing here are not as compelling as for the lower end, and will thus take longer to be worked out and implemented.

Browser Interfaces

One of the powerful aspects of HTML is the way in which it is able to provide a relatively robust interface with a relatively simple, platform-independent language. Although relatively complex GIS interfaces can be created using forms, tables, frames, and clickable graphics, there are some limitations.

Many elements in a standard user interface are dynamic. For example, menus, help, and other elements may appear as you move the mouse to a certain area; one part of a form may change in response to another part, either with new elements or different values; or map images may allow for several clicks, such as when drawing a polygon.

Solution 1: Improved HTML

This problem would be solved if the capabilities of HTML, especially forms, were improved to support these functions. The specifications for this language are constantly under development, and many improvements will likely become available that will allow for better DGI services.

However, this cannot be relied on to solve the problem. Primarily, this is because the development of HTML is entirely out of the control of the DGI community. The desired changes may or may not arrive at some point, but those who are creating services would like to be able to improve them sooner rather than later.

Solution 2: Extensible Viewer Interfaces

Java, ActiveX, and plug-ins remedy this problem due to their more robust interface capabilities. Thus, each service could have an applet supporting its needed interface functionality. However, this is not a desirable solution because of the nuisance factor of having to download a unique applet for every application.

The standard map viewer described in the previous section would solve the nuisance problem. However, if it were implemented in the same way as today's applets, it would have a set interface that would not work well with the variety of DGI applications out there.

A better approach would be to have a viewer to which the server could add new interface tools on the fly, along with the map, similar to the way a map and interface are sent in HTML. The standard graphics language and/or protocol would include commands to generate various interface elements similar to those in HTML forms, but more complete.

Much of the nuisance associated with transferring entire programs over the network (as experienced with Java) could be avoided by "faking" the functionality, the same way that HTML forms are fake functions. That is, operations such as "buffer" and "query" would not be included in the code sent with the map, only buttons and forms for requesting those operations. When the user fills out the request, it is sent back to the server for processing, with another map/interface package being returned.

Essentially, the result would be very similar to how the normal HTML interface works (as described in Chapter 6), but with more complete design capabilities, and much more geared toward a map-based, rather than document-based, interface. It would also have simple map navigation capabilities built in, such as zooming and panning, saving most of the network traffic incurred in today's HTML-based services, such as MapQuest.

This "fake functionality" approach would have the added advantage of access control. This is because you could allow people to use your applications (stored and executed on your server) without actually giving them your applications (which would happen if they were programmed into the applet).

The main drawback is that it would incur more network traffic than if the operations were performed on the user's computer. It would also put more load on your server computer. This possibility has been raised in the early open map sharing discussions in OGC (as described in the previous section), and may be included in any resulting specification.

GeoData Commerce

DGI is potentially a very worthwhile commercial venture for many organizations, because most users would be willing to pay a reasonable amount for valuable information. Fifty cents for a map highlighting the route from the airport to your hotel is trivial for the user, but potentially profitable for the service provider in bulk. However, there are two main obstacles that may prevent many companies and other groups from entering this arena.

Nuisance

One problem is that paying 50 cents every time you get a map is prohibitively inefficient using the currently available methods for Internet commerce. Nobody will be willing to fill out a form to pay this little with a credit card, and only frequent users will pay a monthly fee to subscribe to such a service.

Electronic cash services represent one solution to this, because they allow for rapid, anonymous payments regardless of the amount, essentially being no more cumbersome than traditional currency. Although the technology exists today, these services are not in wide use because they require special accounts and because few sellers (and very few brands of HTTP server software) support them. As banks begin to allow patrons to easily withdraw electronic cash from their existing accounts, this will become much more widespread.

Another, simpler solution, is to still pay with credit cards, but without the cumbersome forms. The necessary information would be stored in the user's browser, and automatically sent to the vendor, as when a product is selected for purchase (with the buyer's approval, of course).

Transaction Security

The other "problem" is transaction security. A large number (by some indications, a majority) of Internet users do not trust the ability of the network to protect their credit card information as it is passed to the server. Nor do they necessarily trust the identity of the server; that is, is *www.sears.com* really Sears Roebuck & Company?

This type of security is actually much easier than a company trying to protect its internal information from hackers. The Secure Sockets Layer (SSL) used in Web commerce servers is at least as safe as giving your credit card number over the telephone, and the server owners are required to prove their identity in order to use SSL, but it can probably be improved. Also, in time (and with more positive media coverage), users will feel more comfortable with this new means of payment.

Isolated Applications

Currently, DGI applications are run in a Web browser, a program entirely separate from the rest of the applications on a user's computer. However, there are many instances in which maps and other data from the Web needs to be integrated with other types of information.

A Typical Scenario

For example, a marketing executive may want to maintain a sales report with data collected from regional offices over the Internet. In addition to statistical data, the executive wants to incorporate a map from the company's mapping Web site, with points showing the locations of sales.

To do this today, the map would be generated in a Web browser, saved to the disk from the Web browser, then imported into the spreadsheet (assuming the spreadsheet software supports the format of the image). This is not too difficult for a one-time inclusion, but if the data is dynamic, this process is prohibitive because the person would need to go through the import process every time the data changed.

Solution: Compound Documents

The answer to the problem of isolated applications is to make it possible for documents to receive updates directly from the Internet without an intermediate Web browser. Thus, the spreadsheet would include live links to the map service, as well as to the data sources. Although current trends in monolithic application software development (such as the enormous suites from Microsoft, Corel, and Lotus) may eventually lead to such capabilities, a better prospect lies in the technology of compound documents.

Rather than being created and maintained by a single multifunction application (i.e., the spreadsheet software), a compound document is composed of pieces generated by many small, single-function programs executed only when necessary. In this scenario, the sales report document would consist of the main spreadsheet section, along with one or more sections controlled by a Web browser program.

One of the Web pieces would be the map image. Because the map image section is under the direct control of the Web browser rather than the spreadsheet, it could be updated as easily as any Web document. In addition, because it is an integral part of the compound document, it could be placed anywhere in the report, even in the middle of the spreadsheet.

Microsoft's Object Linking and Embedding (OLE) is an early attempt at this, but more streamlined approaches such as ActiveX and Java Beans are promised. The major holdup to these being implemented is probably not in the technology but in the fact that they force applications vendors to completely rethink their business models. That is, how can companies such as Corel sell myriad small program components instead of the bulky programs now popular? It is likely that the monolithic suite approach will continue, even if the underlying structure of the software is composed of many small components, including Web linkage components and perhaps even map creation components.

Isolated Data Sources

Another problem occurs with sites that distribute raw GIS data, whether via direct download or online ordering. Although it would not be difficult for you to use the principles in this book to create a server that allowed you to give away or sell your data, it may be difficult for users to find *you*, as described in Chapter 6 under "Promotion." The main problem here involves the multiplicity of information sources.

Multiplicity of Sources

Even if potential data users hear about your site, they may not visit simply because of the multiplicity of similar sites available. An analogous situation would be if every book publisher had its own chain of bookstores and its own chain of libraries.

If a person wanted a book on a particular subject, he or she would be faced with the daunting task of going to numerous stores to research what was available for that topic. Most patrons would simply decide to forget the whole affair and watch television instead. By using a pooled outlet, all publishers are able to make more sales. This approach might be applied to online digital information.

Solution: Digital Supermarkets and Digital Libraries

Fortunately, the technology of the Internet allows us to build a similar pooled outlet without having to physically co-locate all of the terabytes of data available. Clearinghouses and digital libraries—such as those developed by the FGDC (*http://www.fgdc.gov*), Core Software (*http://www.coresw.com*), and the Alexandria Project (*http://alexandria.sdc.ucsb.edu*)—provide a centralized interface that can search "libraries" and "supermarkets" around the world.

It is not certain whether the technologies currently used by these products are scalable. That is, can they search a hundred data servers? A thousand servers? A million? The technology will likely continue to improve to keep up with demand.

There will likely be a demand for pooled data servers as well, because it is not in the best interest of every university department or small business that may have a few data sets to build and maintain its own data server. One can envision "co-op data markets," where several small data producers could place their data sets (or at least the metadata for their data sets) on a common powerful server, contributing a portion of their data sales to the maintenance of the server.

Within an organization, this problem is being addressed through "data warehousing," roughly the same thing as described above, but on an Intranet. A single interface is created for accessing all of the organization's data, whether stored on the central server or around the office.

Other Issues and Policies

Although the problems discussed in this chapter may very well be solved in the next few years, the items that follow are much more difficult. These are largely societal and legal issues that have to do with the conceptions and assumptions with which the various parties involved (i.e., vendors, service providers, and users) approach DGI.

It will take considerable time for these issues to be solved for the industry in general. Fortunately, they do not require global solutions. As individual vendors, organizations, and users resolve these issues to their own satisfaction, these entities can fully enter the DGI arena.

Data Stinginess

Almost anyone has data they do not want distributed outside the organization. Finances, detailed plans, tax forms, source code for new software, and personnel data are generally not the types of things you want on your Web site, at any price, although they will often form an important component of an intranet.

On the other hand, some information is specifically intended for distribution. Government agencies are accustomed to making their data resources available to the public (at a price), and many companies are in the business of producing and selling spatial data.

The Data Sharing Climate

In between these two extremes, there is a great deal of information that is neither secret nor public. It is then up to the data producer to decide whether to release the information or not.

However, more often than not, the data is "left in the vault." This is not necessarily because the producers are being stingy, or feel their data should be kept secret, but is often simply because the organization is not in a frame of mind to encourage data sharing in principle, even within its own walls.

Solution: A Data Sharing Mindset

Despite the motivations for data sharing discussed in Chapter 1, it may require a complete paradigm shift for an organization to see that the data it has lying around is potentially an easy source of revenue. For example, if a civil engineering firm is creating a site plan for a new development, it may obtain aerial photography and produce an orthophoto of the surrounding area.

Although the plan itself is the property of the developer and should not be released (in its original detail at least), the orthophoto may be worth several thousand dollars to a developer of an adjacent property. This person would otherwise have to spend much more than that to pay for a fly-over of the site to produce the same orthophoto. It takes a new mindset on the part of the firm to realize that there is nothing inherently private or secret about the photo itself, and that making it available (at an appropriate cost) to other organizations, even competitors, is a positive (i.e., profitable) move in the long run.

Another route for those still concerned with keeping their own data to themselves is to distribute corrupted data. This does not mean the information is bad, but that it is not as detailed as the original. For example, the civil engineering firm may sell a version of the orthophoto with a greatly reduced resolution (3 meters instead of 3 centimeters). Thus, casual users such as people living in the area could purchase a pretty picture, and it may be well suited to many research and municipal applications, but it would not be accurate enough for the firm's competitors to use it to outbid them on a nearby project.

Conflation

One of the most promising goals of DGI is to enable users to gather data from a variety of sources. Both the net-savvy GIS and the standardized map viewer (discussed earlier in this chapter) hold the promise of conflating (combining) themes from very disparate producers, or even from separate servers on an intranet.

Types of Conflation

There are two types of conflation. In the first type, multiple themes representing various aspects of reality are combined through an analysis method to investigate their relationships. For example, temperature and precipitation themes may be combined to yield a climate theme.

The second type of conflation is less common and more complex. Here, themes representing the same aspect of reality are physically combined to yield a single compromise representation. For example, by averaging the geometry of each road segment and extracting the attributes for that segment from both themes, a road layer from U.S. Census TIGER files might be combined with roads from USGS DLG data sets to yield a single road theme.

Common Conflation Problems

Although either method of conflation can be very useful, it can also cause serious problems for many reasons, most resulting from the fact that the themes come from different sources. Because no data set is perfectly accurate, and because different data entry approaches may yield different results, it is highly likely that the input themes will have different levels of accuracy.

If the accuracy of each theme is taken into consideration (e.g., weighting each decision according to the reliability of each input), the resulting themes will usually be just as accurate as the original themes. However, if theme accuracies are not considered, and all input themes are assumed to be accurate (or at least at the same level of inaccuracy), more incorrect results will follow.

There is also a potential semantic problem, especially with categorized data. For example, the highways from one theme may not match well with the highways from another, because the producer organizations may have different definitions of the term *highway*, and thus different ideas about which road segments should be included in that category.

Another problem has to do with the geodetic base (i.e., projection, coordinate system, datum, or spheroid model) of each theme being different. Although projections and coordinate systems are usually easy to recognize and match, the datum and spheroid are much more subtle. However, if they do not match, neither will the geometry of the two themes, by up to several hundred meters or more.

Solution: Use of Metadata

All of these problems regarding conflation, and other issues, can probably be solved (or at least avoided) through the use of metadata, which describes the characteristics of each data set in detail. If users read the metadata records for themes being used, they are aware of what must be done to make them match. However, only the most intrepid users will be willing to read and interpret several complete metadata records, because the records are so large and complex.

One answer to this problem is to add capabilities to the GIS software or map viewers to automatically read and interpret quantitative metadata associated with each data theme used. The software would then be able to alert the user when a potentially inappropriate operation was requested (which could be overridden, of course, if the user knew what he or she were doing). The software might also contain tutorials to guide the data user through the process of making the input data sets match, or even perform that process automatically.

Software and Service Pricing

One of the most difficult decisions to be made in starting a commercial venture in a new field is how much to charge for services. This can be a complicated assessment because the demand side of the economic equation is largely unknown, as is the sensitivity of that demand to changes in price.

Thus, when any industry is in its infancy, the prices of the products often fluctuate rapidly but are usually extremely high, so that just a few sales will offset the cost incurred during development. This has been the case with GIS software. In the early days, only large government and military installations used GIS, and the prices were very high, effectively shutting smaller potential users out of the market. As it has become more popular, and with the competition of desktop mapping software, the price has decreased greatly while allowing vendors to retain a profit margin.

This may help explain the relatively high price currently charged for commercial software, and help you decide on a price to charge for the use of your information. However, it is important to determine a reasonable price as well as possible at the outset, using a method such as market surveys. If you continually change your prices according to shifts in demand, users who paid for your information when the price was high will not be pleased.

Another reason for the high price of some commercial software has to do with their business model. Generally, vendors are able to sell GIS software directly to users for a per-copy price. However, when you build a DGI service (especially one with extensive analysis capabilities), you are effectively decreasing their market by providing your own software to those users.

Vendors' prices often reflect that fact. In essence, they are selling you many copies of the GIS software instead of selling them to your users. In many cases, it is completely permissible for you to pass this cost on to those users, especially if they are getting a great deal of value from your services.

Intellectual Property

When considering the Internet, one of the major concerns of information producers (with the exception of most U.S. government agencies) is the protection of intellectual property rights. How can you distribute data you have produced but prevent customers from violating your copyright? One might first ask exactly what is a violation and what is not.

This issue has spurred a great deal of debate in the Internet and legal communities. In fact, new legislation has been proposed in many countries in an attempt to deal with this conundrum. However, in at least some countries, most Internet situations are adequately covered by current intellectual property law.

Fair and Unfair

Fair use is one of the legal concepts in use today that can be effectively applied to the Internet and DGI. This principle (at least in the United States) basically says that if someone obtains a copyrighted product legally (i.e., with the copyright owner's permission), he or she can make copies legally for personal use. If that person begins to redistribute your information (i.e., stealing other potential customers), it is considered a violation of copyright. This is true even if the information is augmented or altered, or if they give it away for free.

Therefore, if you have a map browsing service, when a user views the page, the map image is downloaded, and a copy resides on the user's disk. This is fair use, whether they soon delete it, keep the copy around for future reference, or even print it out. However, if that user places the page on his or her own Web page (e.g., "Here is my house") without your permission, it is probably in violation.

Partial Solutions

The truth is, you cannot prevent illegal copying using technology. There is probably no foolproof technical trick that will make it impossible for people who buy your data to use it inappropriately. This is not really possible with any type of intellectual property. However, there are several workarounds that should discourage most users from breaking the law.

The first is the "watermark." As with its historic namesake, this is a method of embedding an identifying mark within your information, which cannot be removed, and if well done, is not recognizable to anyone but yourself.

There are several techniques for creating a watermark. One that has been used for a long time by cartographers is to include false but insignificant features such as misspelled towns or extra meanders in a river. In a more technical approach, Adobe has recently developed a method for encoding messages into the bits that constitute a raster image, which will usually be preserved even if the image is altered. Similar techniques could be used for vector graphics as well.

Watermarks are used in two ways. First, they can be used as evidence if legal action is taken against the violator. Also, if you make it known that you are using watermarks (but not exactly how), you can deter many would-be thieves.

The second method is the same as that suggested in the data sharing section. If the data you distribute is to be used for viewing only (i.e., you do not want people to be able to input it into their GIS to make new products), you can degrade the data sufficiently to make it useful for viewing but not much else.

For example, you could display a raster map in GIF or JPEG format; they look nice, but they are not much use without extensive metadata on the projection you use to generate the pixel coordinates. If you are distributing vector maps, you can do the same thing by drawing the maps in a false coordinate system such as map inches instead of real-world meters. It will look fine on their screen, but will not be positioned correctly in a GIS.

Of course, truly intrepid users could manually re-register each of your maps to real-world coordinates and piece together many maps to recreate a crude approximation of your original data set, but this would be extremely difficult.

Another approach is to capitalize on your users' desire to copy your data. Many data producers offer reseller licenses, selling the rights to alter and/or redistribute the data and maps, with appropriate restrictions and for an appropriate price. Although many users will still copy your information illegally, there are a greater number of honest users and organizations who are willing to pay to do things legally.

Postscript

This book is intended to help guide you through the process of creating services that use Internet technologies to distribute geographic information, including spatial data sets, maps, and analysis operations. This process includes three major phases: planning your service (discussed in chapters 2 through 6), building the service (discussed in Chapter 7), and permanently maintaining and augmenting the service (also discussed in Chapter 7).

Each organization that might use this book is unique, and each service created is unique. There is also a wide variety of techniques and software available to accomplish any design. Thus, this book has not attempted to give detailed step-by-step instructions on how to accomplish each phase of the process.

Instead, it discussed the issues and options you should consider during the process, helping you make decisions about the approach you will take. At each stage, the variety of solutions available have been presented in a common framework, allowing you to compare them directly. By taking a broad perspective, it is hoped that you can make informed decisions, albeit difficult ones.

Because detailed instructions are not provided, you will need to use other sources more tailored to your particular solution. The user manuals provided with the HTTP, DGI, and GIS software you decide to use (at least if they are commercial solutions) will be your most valuable resource. Each vendor's technical support (whether by telephone, fax, or online) will also be invaluable at times.

Last, you should find existing services similar to your own design, and consult them for hints and ideas. Places to look for DGI services are listed in Chapter 6 under "Promotion."

This book starts you on the road. As you follow the principles outlined herein, the services you create will be attractive and useful to your intended audience, and will help your organization further its goals, whether for service, for profit, or for both.

An Illustrated Glossary of Internet and GIS Terminology

Note: This glossary is included as a resource to introduce many of the concepts of geographic information systems and the Internet, focusing on those terms used frequently in the book. Where applicable, the URL addresses of Internet sites that explain the concept in more detail are included. Italicized terms within the definitions have their own entries in the glossary. See also the notes on the glossary in the introduction to this book.

Backbone. When people travel across a country, they generally do not use minor streets, used for local traffic, but funnel onto major highways or express rail lines. In the same way, Internet traffic moving over large distances is funneled into a system of very high bandwidth lines known as a backbone.

The Internet is a conglomeration of several regional and commercial backbones loosely interconnected. When you send a message to a remote computer, it moves from your computer to your ISP, then to a backbone, then possibly to another backbone, then to the other computer's ISP, then to your destination. The ANSNet backbone is shown in the following illustration.

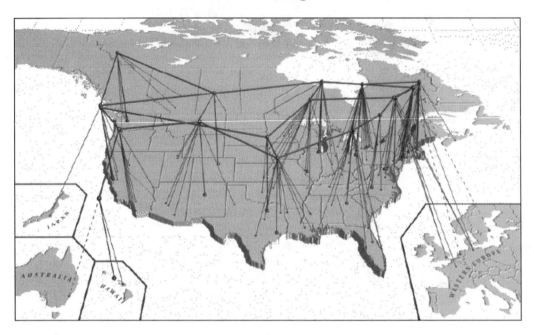

A representation of ANSNet, one of many Internet backbones.

Bandwidth. This term refers to the rate at which data can be sent over a connection between two computers on a network. As with any flow, bandwidth is measured as the volume of data passed in a given amount of time, expressed either in kilobits per second (Kbps or "kilobaud"), megabits per second (Mbps), or megabytes per second (Mbps). A high bandwidth line, such as a T3 (45 MBps), can send larger files in less time than a low bandwidth line, such as a 28.8-Kbps modem.

As the Internet grows (i.e., more computers trying to send data over the same line simultaneously), and becomes more complex (i.e., multimedia objects such as video and maps constitute much larger files than text), the need for bandwidth is constantly increasing. Whether the Internet can be upgraded fast enough to keep up with demand is uncertain.

Browser. *See* Web browser.

Browser plug-in. A small program that can be attached to a *web browser* to enhance its capabilities. The most common plug-ins are used to view specialized types of data such as video, animations, and complex graphics (such as maps). Each plug-in is registered with the browser for certain types of information. When items of those types of information are requested by the user, the plug-in is activated to handle them.

Plug-ins must be downloaded from the developer, and installed outside the browser. Because of this inconvenience, people tend to access sites that require plug-ins they use frequently, and avoid sites that use plug-ins they do not have.

Client. The side of a *client/server* application that provides the user with a means of accessing information and applications resident on a *server*. A *Web browser* is one example of a client. The term is sometimes used for the user's computer, but generally it refers to a specialized program that communicates with the server over the network. The client software has a visual interface that allows the user to enter requests (e.g., when you click on a link on a Web page), passing them to the server for processing. When results are received, the client displays the information to the user.

Client/server model. A common architecture (shown in the following illustration) for distributed computing applications, such as the Internet or a large GIS installation. Information and applications are stored on one or more *servers*, which can be accessed by any number of *clients* on computers connected to a network.

The model generally works in a question-and-answer fashion. The user enters requests or commands into the client, which sends them to the server. The server uses its stored information to answer the request, returning information to the client, which displays it to the user.

The client/server architecture.

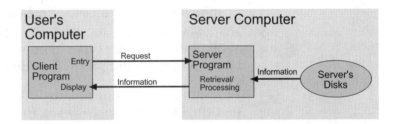

Common Gateway Interface (CGI). A *protocol* for adding specialized processing capabilities to a *Web server*. A CGI script is an independent program that processes data based on certain parameters. When the browser makes a request to the server, the server recognizes the request as the domain of a particular CGI script. The parameters are passed, using the CGI format, to the script for processing. Its results are then passed back to the server, which sends the information to the browser.

Several commercial GIR programs, such as GeoMedia's WebMap, are implemented in CGI. The WWW architecture is shown in the following illustration. The architectures for ISAPI and NSAPI modules are roughly the same.

The WWW architecture augmented by a CGI script.

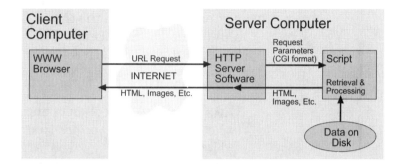

Computer Graphics Metafile (CGM). An international standard *data format* for *vector* graphics. Although not as full-featured as the native formats of most graphics software (i.e., it cannot handle complex graphic effects, and the text fonts are usually less than desirable), it can work for displaying most styles of vector-based maps. A Web-aware extension, known as ActiveCGM, has been developed by InterCAP, which can attach URLs to each graphic element. This enables many interactive mapping applications.

Coverage. The term used in the ARC/INFO GIS platform for a *data theme.*

Data format. In order for any information (including geographic information) to be usable by a computer program, it must be digitally encoded in a way the program can understand. The specific structure in which it is encoded is called a data structure, data format, or file format. In the case of *vector*-like geographic information, data formats include the order and style in which the coordinates representing each object are written, as well as any attributes of that object.

Data layer. *See* Data theme.

Data set. Similar to a *data theme* in that it is a complete group of information (geographic in this context) describing a particular theme. However, this term is normally used when the theme has been packaged into a unit, such as a *DLG* or *DEM* file, designed for distribution outside the GIS software.

Data structure. *See* Data format.

Data theme. A collection of one or more digital objects (whether *raster* or *vector*) that describe a particular subject matter for a given area. For example, in a municipal GIS, there might be several separate data themes (see the following illustration), representing roads, parcels, city limits, water bodies, terrain, earthquake hazard zones, and so on. Each theme may consist of many objects representing individual geographic entities, but they are stored as a unit—either a single file or a small group of files.

Several of the data themes that may be part of a GIS.

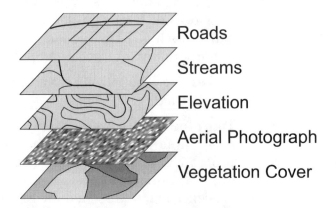

Roads

Streams

Elevation

Aerial Photograph

Vegetation Cover

Digital elevation model (DEM). *(a)* Any *data theme* that represents terrain (the shape of the earth's surface); the most common structures being a *raster* grid of elevations, and the *vector* triangulated irregular network (TIN). *(b)* The particular raster *data format* used by the USGS for distributing its terrain *data sets*.

```
http://edcwww.cr.usgs.gov/nsdi/gendem.htm
```

Digital line graph (DLG). A series of *data sets*, created by the USGS, that consist of *vector* cartographic themes (e.g., roads, rivers, and boundaries) digitized from their topographic maps. The term is also used for the *data format* in which these data sets have traditionally been distributed, although the USGS is abandoning that format in favor of *SDTS*.

```
http://edcwww.cr.usgs.gov/nsdi/gendlg.htm
```

Digital orthophotoquad (DOQ). A *raster data theme*, built from scanned aerial photography, rectified and mosaicked to form a planimetrically correct map-like product. Used as a base for digitizing features, or as a realistic backdrop for maps.

Distributed file system (dFS). A recent technology built into current versions of Windows 95 and NT, which enables access to disk drives on remote computers over the Internet in the same manner as local disks and the local area network. Essentially, dFS is a similar product to *NFS*, but more Windows-like than Unix-like.

```
http://www.microsoft.com/ntserver/info/dfswp.htm
```

Domain Naming Service (DNS). A standard program on the Internet that allows you to use an alphanumeric name (called a "host name") for a computer in lieu of the numeric *IP address*. DNS names each machine in terms of its top-level domain (e.g., .com, .edu, .jp, and .de), organization domain that mimics the name of the organization (e.g., byu and texol), optional subnetwork designator representing a department or office (e.g., sales), and machine name (e.g., tomlin).

Combined with periods, host names look like "tomlin.sales.texol.com." In the common host name configuration of www.xxx.com, www is the name of the machine, whereas xxx.com is a domain name common to all of the computers the "xxx" organization has on the Internet. It is possible for a machine to have multiple names, either as aliases for the same IP address, or as unique assignations for multiple addresses to the same machine. The latter approach is how *ISP*s can host *Web servers* for many organizations with unique domain names.

Drawing Exchange Format (DXF). A *vector data format* developed by Autodesk. DXF is commonly used for transferring cartographic data between software platforms, but seldom used for complete GIS data transfers because it does not handle thematic attributes.

File Transfer Protocol (FTP). An Internet *protocol* commonly used for downloading and uploading files from a server. Users log in (UNIX-style) using an allowed user name and password, although many servers allow "anonymous" access using a guest user name. Protocol commands are similar to Unix file management commands, but most people use specialized GUI programs, or gain access through a web browser (which can usually speak FTP as well as HTTP).

The advantage FTP has over HTTP is that it has *state*, meaning that the client-server connection stays open between transfers. This makes sessions of many transfers operate relatively quickly, although FTP single transfers are slower than HTTP transfers because of log-in overhead.

Firewall. A security tool consisting of a computer that sits at the connection between an organization's local network and the Internet. The firewall server controls all communications passing out to the Internet and back into the organization. Its intent is to allow personnel access to the Internet (for e-mail, Web uses, and so on); however, it bars the Internet at large from accessing or modifying sensitive information stored on the internal network. The organization's public Web server is usually placed "outside" (on the Internet side) the firewall so that it can be accessible.

Geographic information science. The academic field of study concerning the issues and fundamental concepts behind GIS. Common topics include the design of effective and efficient data structures, intuitive user interfaces, issues of accuracy, and the formal description of geographic entities, including their philosophical nature. Outside the United States, the term "geomatics" is often used for this field to avoid confusion with the more applied usage of the acronym GIS, *geographic information systems*.

Geographic information system (GIS). This term is used interchangeably to refer to three distinct, but related, entities. *(a)* A digital database that includes the geometric form of spatial entities (e.g., a string of coordinates that outline a lake), as well as thematic attributes (e.g., for a lake, the maximum depth, volume, surface elevation, salinity, and pH). *(b)* Software (such as ESRI's ARC/INFO and Intergraph's MGE) used to manage, analyze, and display the database of definition *a*. *(c)* A complete installation at an organization, consisting of the necessary computers, software, and database for a given application.

Graphics Interchange Format (GIF). A common *raster* graphics *data format* used on the Internet to display images. Because of the method it uses to compress raster data, it works well (storing the same information in very few bytes) for simple graphics with up to 256 colors. However, for complex images such as photographs, *JPEG* is more useful. Unlike JPEG, the compression technology used in GIF (the Lempel-Ziv-Welch, or LZW, algorithm) is owned and copyrighted by Unisys, and data/software producers (but not users) must pay a licensing fee to use it.

Host. *See* Server.

Host name. *See* Domain Naming Service; IP Address.

HTTP server. The computer and software that contains information to be distributed over the W*orld Wide Web*. The information may be documents and other static files (such as *HTML* pages), databases that can be queried, or any other type of service. The server receives requests from remote *Web browsers*, which communicate using the *HTTP* protocol.

After each request is processed, the results are returned to the client, and the HTTP session is ended. The specific requirements for an HTTP server computer vary, but it must be powerful enough to process many requests simultaneously, without slowing down excessively. There is a variety of commercial and free options for HTTP server software, with various capabilities for delivering documents and other services.

Hypertext Markup Language (HTML). The *data format* used for textual documents on the *World Wide Web*. A derivative of *SGML*, HTML uses a raw text file to store the text of the document. The text is interspersed with codes called "tags," which control the action and display of the document. These tags include such things as headings, paragraph formatting, tables, and inclusion of other files in the document such as graphic images (which are downloaded in *HTTP* requests separate from the original HTML document).

HTML is designed to be *platform independent* so that browsers on any computer (powerful or not) can make sense of the content. However, some extensions are more platform specific, understandable by only a few browsers on some computers.

```
http://www.w3.org/pub/WWW/MarkUp
```

Hypertext Transfer Protocol (HTTP). The Internet *protocol* that, along with HTML, forms the foundation of the *World Wide Web*. HTTP is a very simple language patterned after the *client/server mode*. That is, *clients* called *Web browsers* usually request information using a *URL* and other information, which is delivered by an *HTTP server*.

HTTP is *stateless*, meaning that as soon as a request-response communication is complete, the connection is broken. If a user clicks on a hypertext link to get a new page from the same server, a new connection is formed, with no memory of the last one. This is very efficient for the basic browsing common on the Web. However, it can be quite cumbersome for the popular interactive services such as GIR because these services need to remember what the user has been doing up to and including the most recent request. Future versions of HTTP will include an option to leave connections open throughout a session.

```
http://www.w3.org/pub/WWW/Protocols
```

Internet. Probably the most successful anarchy ever, the Internet consists of millions of computers of various types, physically connected to one another by lines of many types owned by many companies and communicating via several languages (called *protocols*). Often referred to as "a kludge that works," no single organization owns or controls the entire network, although it is generally controlled by a series of accepted standards. The *World Wide Web* is only one form of communication that takes place over the Internet, alongside other forms such as electronic mail.

Internet Server Application Programming Interface (ISAPI). Pronounced "I-za-pee," this is a method for adding processing functionality to an *HTTP server*, analogous to CGI and NSAPI (although much faster and more powerful than CGI because it is more tightly integrated with the HTTP server software). Developed by Microsoft as part of their IIS server software, it is now supported by many server platforms.

As with the other server extension interfaces (as shown in the illustration under the entry "Common gateway interface"), it is activated when the HTTP server recognizes a request as belonging to its domain (usually by a flag in the *URL*, such as a unique file name extension). Results of its processing are passed back to the server, and subsequently to the requesting browser.

Internet service provider (ISP). A company that provides connections to the *Internet* to individuals and organizations for a fee. An ISP has its own high-*bandwidth* connection to an Internet *backbone*, and allows users to access this connection using either dial-up modems or, for faster access, direct lines, which are leased to an organization. There are nationwide ISPs, such as Sprint and MCI, in most countries, as well as locally owned businesses in almost any city. An ISP connection is shown in the following illustration.

The connection between an organization and the Internet via an ISP. The width of lines is proportional to bandwidth.

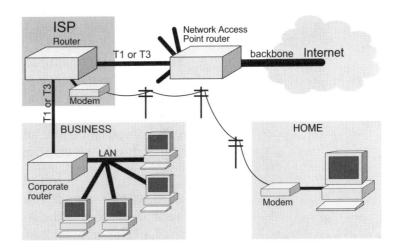

Intranet. A system similar to the *Internet* but used to communicate only within an organization. It uses standard Internet *protocols*—such as *HTTP*, SMTP, and *TCP/IP*—and is based on the same distributed *client/server* idea, with many *server* computers, and *clients* on every desktop. However, everything is located on the local area network, and is usually barred from the rest of the Internet by a *firewall*. It is commonly used to share information used by many members of the organization, such as finances, plans, reports, staff information, and announcements.

IP address. In order to facilitate communication between computers on the *Internet*, each computer is assigned a unique address. This consists of a four-byte number, usually written as each byte separately with periods in between, such as 200.56.24.100. Using the *Domain Naming Service*, an alphanumeric name can be assigned to each address, such as bill.hero.com, which is much easier to remember.

This four-byte system allows for only 4,294,967,296 computers, the limit of which is in danger of being reached soon. To get around this, schemes are used such as dynamic address allocation, in which your computer does not have its own address but is assigned one from a small pool only when you connect to the Internet. IP-NG (Next Generation), currently under development, promises a new addressing system that will allow for a much larger number of computers.

Java. An increasingly popular object-oriented programming language for developing Internet-based applications, both as part of the *Web* and as independent applications. Developed by Sun Microsystems, its power lies in its *platform independence* (the same concept that makes *HTML* so useful on the eclectic Internet).

Because the language is interpreted (like *Perl*), it is not tied to a specific CPU, and can be run on any computer with an interpreter (called a Java Virtual Machine). Therefore, specialized programs (such as custom GIR interfaces) do not need to be purchased and installed but can be downloaded and run whenever needed. Another advantage is that it is safe for wide distribution because downloaded programs cannot have access to your local disk and configuration.

```
http://www.javasoft.com
```

JavaScript. A derivative of *Java* (developed by Netscape) used to add simple functionality to *HTML* documents and *HTTP servers*.

```
http://home.netscape.com/eng/mozilla/Gold/
handbook/javascript/introd.html
```

Joint Photographic Experts Group (JPEG). An industry committee that develops standards for digital photography. The acronym is commonly used for the raster *data format* this group created, which is more accurately called the JPEG File Interchange Format (JFIF). JFIF files are compressed in a fashion optimized for true-color images such as photographs and *remote sensing* imagery, and thus store them much more efficiently than *GIF*. The major drawback to JFIF, however, is that it is "lossy." That is, a small amount of information is lost in the compression process, making a JFIF image of slightly lower quality than the original image.

```
http://www.faqs.org/faqs/jpeg-faq
```

Metadata. Literally, "data about data," this is information that describes the general characteristics of a *data set*. For geographic data, metadata commonly includes specifications of the *data format*, the projection and coordinate system, the spatial area covered by the data, the accuracy of contained locations and attributes, data lineage (the original source of the data), thematic subjects included in data, and information about how to obtain the data set itself.

Metadata is vital in helping potential users determine whether a data set will meet their needs before they spend the time and money to obtain and process it. In the United States, FGDC has developed a standard structure for geospatial metadata, which is the basis for the international (ISO) standard currently being developed.

```
http://www.fgdc.gov/Metadata/metahome.html
```

Netscape Server Application Programming Interface (NSAPI). A method for extending the functionality of an *HTTP server.* This interface was developed by Netscape for its server software (although many other servers now support the protocol). Fulfilling the same purpose as *ISAPI* as a successor to *CGI,* it allows site developers to create specialized applications used on demand by the HTTP server when a client requests that functionality. Many of the DGI programs described in Chapter 4 are implemented as NSAPI modules.

```
http://home.netscape.com/misc/developer/
conference/proceedings/s5
```

Network File System (NFS). An Internet *protocol* for sharing files on a network. This system is similar to the standard Unix file system, except that disks on remote computers (on the local network or on the Internet) can be mounted locally. Access to a shared drive can be made open to everyone, or limited to authorized users. NFS client and server software is an integral part of most Unix systems, but drivers are available for other operating systems.

```
http://www.sun.com/sunsoft/solaris/desktop/nfs.html
```

Platform independence. One of the best points about the Internet is that it connects a great variety of computers, from mainframes to laptops. These machines have a wide range of processing speeds, display capabilities (e.g., colors, fonts, and resolution), and network connections. This diversity also forms an obstacle, because there is a need to design information and applications that can be used on all of these computers.

Many Internet-based systems—such as *HTML* and *Java*—and most *protocols* are designed to be independent of the platform used. They are relatively simple in structure (e.g., using text instead of binary codes), often designed in open committee processes.

Plug-in. *See* Browser plug-in.

Portable Network Graphics (PNG). A raster *data format* similar to *GIF*. It is an open standard developed in response to the restriction of the copyright on the GIF compression.

```
http://www.wco.com/~png
```

Practical Extraction and Report Language (Perl). A programming language originally designed for automating UNIX system administration. Because it is a high-level interpreted language, and especially adept at manipulating text files such as *HTML*, it has been popular for writing simple *CGI* scripts for the Internet. Although the most recent version of Perl is object-oriented, Java has become more common for performing complex tasks.

```
http://www.cis.ufl.edu/perl
```

Protocol. On the Internet, a protocol is a language by which two computers communicate. Common protocols include *HTTP* (used on the Web), *FTP* (for file downloading), SMTP (for e-mail), and *TCP/IP*. A protocol normally consists of a standard format for issuing requests and commands, as well as delivering prescribed forms of information.

During a communication, many protocols may be used simultaneously on different levels. For example, when you retrieve a page on the *Web*, the *HTTP* protocol is used to format the file so that the browser can understand what to do with it. However, *TCP* is used to actually manage the file transfer, and at an even lower level, *IP* controls the physical transmission of the data from one computer to another.

Raster structure. One form of organization for spatial data (whether geographic, photographic, or just graphic). The area being represented is divided into a rectangular array of regularly spaced squares called cells or pixels. Each cell can be assigned a single attribute, such as the color to be displayed or the average soil moisture in the cell area.

Raster files can be quite efficient for representing complex field-like *data themes* of continuous change, such as a temperature signature or a photographic image. However, for more simple, geometric entities such as roads or text, *vector* structures usually work better. Printers and monitors are common raster output devices. Raster *data formats* include *GIF*, *JPEG*, and *DEM*. Adobe Photoshop is an example of raster graphic software. Raster GIS programs include GRASS, IDRISI, Tydac SPANS, and ArcView Spatial Analyst. The following illustration shows a raster data structure.

A raster data structure.

Soil pH

Coordinates in State Plane feet

	123400	123450	123500	123550	123600	123650	123700
65400	7.6 7.7 7.8 7.6	7.4 7.3 7.1 7.0	7.2 7.2 7.0 6.8				
65350	7.5 7.7 7.9 7.7	7.4 7.1 7.0 6.9	7.0 7.0 6.9 6.7				
	7.2 7.5 7.7 7.4	7.3 7.2 7.0 6.8	6.9 6.9 6.7 6.6				
65300	7.0 7.2 7.4 7.2	7.2 7.0 6.9 6.7	6.8 6.8 6.6 6.6				
	6.8 7.0 7.1 6.9	6.9 6.7 6.8 6.6	6.7 6.6 6.5 6.4				
65250	6.5 6.7 6.8 6.7	6.6 6.6 6.5 6.4	6.4 6.4 6.3 6.3				
	6.3 6.5 6.6 6.5	6.4 6.3 6.3 6.3	6.2 6.1 6.1 6.2				
65200	6.3 6.3 6.4 6.3	6.2 6.1 6.0 5.9	6.2 6.2 6.3 6.4				
	6.6 6.6 6.7 6.5	6.4 6.3 6.2 6.2	6.4 6.4 6.5 6.7				
65150	7.0 6.9 7.0 6.8	6.6 6.5 6.3 6.4	6.6 6.7 6.7 6.9				

Remote sensing. The use of remote (airborne, space-borne, and sometimes raised platform) equipment to view and study the earth. There are two common products used in GIS: aerial photography and its derivative products—such as *digital orthophotoquads*—and digital satellite imagery, such as that from Landsat and SPOT.

Spatial Data Transfer Standard (SDTS). A U.S. Federal Information Processing Standard (FIPS 173), which is more than a *data format*. It is a standard method for defining many new data formats, called profiles. It defines a common set of terminology, appropriate themes, and partial structures. When the acronym is used today to refer to a specific data format (such as in the *data sets* now distributed by the USGS), it usually refers to the Topological Vector Profile (TVP), the vector data format defined under this standard.

```
http://mcmcweb.er.usgs.gov/sdts
```

Server. The half of the *client/server* model responsible for housing a large amount of information (e.g., documents, databases, and GIS), and for distributing that information to *clients*. Unfortunately, this term is used interchangeably to refer to the computer that contains the information and the software (which processes requests) on that computer. To avoid confusion, use the term *server computer* or *host* for the former entity, and *server software* for the latter.

Standard Generalized Markup Language (SGML). An international standard, SGML is similar to SDTS in that it is not a *data format* but a standard method of defining formats (called document type definitions, or DTDs) for structured text. The structure and display of a document or other piece of text is controlled by codes (called markup), such as "<heading>." *HTML* and the FGDC standard for *metadata* are both implemented as SGML DTDs.

```
http://www.sgmlopen.org
```

Transfer Control Protocol/Internet Protocol (TCP/IP). A pair of integrated *protocols* that form the basis for almost all communication on the Internet. For any computer to be connected to the Internet, it must not only have a physical wire connection but a driver that allows it to speak TCP/IP.

Neither protocol has much to do with the particular type of information being transmitted, nor the applications sending and receiving the information. They are responsible for packaging the data and managing its routing to the prescribed destination. Although these protocols have been built into UNIX for many years, it was in 1995—when they began to be integrated into desktop operating systems such as Windows and MacOS—that the Internet began to skyrocket in popularity.

Uniform resource locator (URL). Because objects on the *Web*—including documents, images, and other services—are scattered worldwide on thousands of computers, a consistent way of locating them is necessary. The URL is a standard format for giving each object a unique address, and is used to point to anything on the *Web*, as well as to other Internet services such as *FTP*.

Although variations in format are common, a URL consists of four basic parts: the *protocol*, such as http, ftp, or mailto; the server computer's *IP address* or name, such as //www.hmp.com/; the directory on the server in which the object is found (in Unix format), such as /books/; and the name of the object itself (usually a file name), such as gisonline.html. Together, these parts would be written as follows:

```
http://www.hmp.com/books/gisonline.html
```

Vector structure. One of the two major data structures for representing spatial information (such as maps and graphics); the other being *raster*. The vector structure is efficient for representing object-like entities, such as roads, buildings, text, and geometric shapes. Each entity is stored as a distinct unit, delineated by a series of coordinates, and has one or more attributes attached, usually in a related table.

Although vector structures, such as the one shown in the following illustration, are predominant in GIS, architecture, engineering, and much of graphic design, *Web browsers* currently do not support them directly. *Plug-ins* or *Java* programs are required to view them online. Common vector graphic *data formats* include

CGM, DXF, and PostScript. Adobe Illustrator, Macromedia Freehand, and Corel Draw are examples of vector graphics programs, whereas ESRI's ARC/INFO, Intergraph's MGE, and MapInfo are popular vector GIS platforms.

A vector data structure.

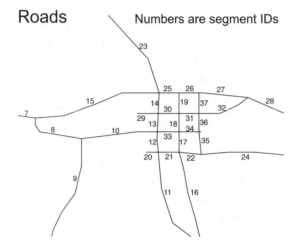

Virtual Reality Modeling Language (VRML). A standard *data format* for describing 3D models to be viewed online using virtual reality software (such as the Live3D module of Netscape Communicator). Although not used extensively for GIS themes yet, VRML has a great deal of potential for visualizing terrain, urban spaces, and global models.

`http://vag.vrml.org`

Web browser. A software program on the *client* side of the *Web* model. Netscape Communicator and Microsoft Internet Explorer are the most common examples. The user has this program on his or her own computer, and uses it to access remote servers using the *HTTP* protocol. Today's browsers are used to not only view *HTML* documents but as general-purpose Internet client tools, with capabilities for e-mail, *FTP*, *VRML*, chat, and NNTP discussion groups.

Web server. *See* HTTP server.

World Wide Web (WWW). One form of communication on the *Internet*, alongside systems such as *FTP*, e-mail, and chat. Although based on the foundation of the *HTML data format* and the *HTTP protocol*, it has grown to include many other types of information, such as graphics and *Java* applications.

Conceptually, it is based largely around the concept of hypertext, where most types of information are portrayed as documents. Certain parts of each document (including text and graphics) have embedded pointers to other documents. Therefore, users can browse the information by following these links. The architecture of a Web client and server is shown in the following illustration.

Architecture of a Web client and server.

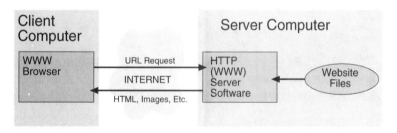

Index

Also Available from OnWord Press

GIS Data Conversion: Strategies, Techniques and Management

Pat Hohl, Ed.

An in depth orientation to issues involved in GIS data conversion projects, ranging from understanding and locating data, through selecting conversion and input methods, documenting processes, and safeguarding data quality.

Order Number: 1-56690-175-8

432 pages, 7" x 9" softcover

The GIS Book, 4th Edition

George B. Korte, P.E.

Proven through three highly praised editions, this completely revised and greatly expanded resource is for anyone who needs to understand what a geographic information system is, how it applies to their profession, and what it can do. New and updated topics include trends toward CAD/GIS convergence, the growing field of systems developers, and the latest changes in the GIS landscape.

Order number 1-56690-127-8

440 pages, 7" x 9" softcover

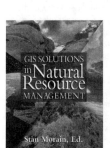

GIS Solutions in Natural Resource Management

Stan Morain

This book outlines the diverse uses of GIS in natural resource management and explores how various data sets are applied to specific areas of study. Case studies illustrate how social and life scientists combine efforts to solve social and political challenges, such as protecting endangered species, preventing famine, managing water and land use, transporting toxic materials, and even locating scenic trails.

Order number 1-56690-146-4

400 pages, 7" x 9" softcover

Raster Imagery in Geographic Information Systems

Stan Morain and
Shirley López Baros, editors

This book describes raster data structures and applications. It is a practical guide to how raster imagery is collected, processed, incorporated, and analyzed in vector GIS applications, and includes over 50 case studies using raster imagery in diverse activities.

Order number 1-56690-097-2

560 pages, 7" x 9" softcover

INSIDE ARC/INFO, Revised Edition

Michael Zeiler

Zeiler introduces the user to the software's basic modules and the commands required to create specific applications. Written to Revision 7.1, this edition includes additional files on a companion CD-ROM and installation instructions for the Windows NT platform.

Order number 1-56690-111-1

680 pages, 7" x 9" softcover

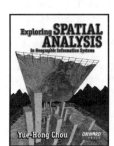

Exploring Spatial Analysis

Yue-Hong Chou

Written for geographic information systems (GIS) professionals and students, this book provides an introduction to spatial analysis concepts and applications. It includes numerous examples, exercises, and illustrations.

Order number 1-56690-119-7

496 pages, 7" x 9" softcover

Processing Digital Images in GIS: A Tutorial Featuring ArcView and ARC/INFO

David L. Verbyla and Kang-tsung (Karl) Chang

This book is a tutorial on becoming proficient with the use of image data in projects using geographical information systems (GIS). The book's practical, hands-on approach facilitates rapid learning of how to process remotely sensed images, digital orthophotos, digital elevation models, and scanned maps, and how to integrate them with points, lines, and polygon themes. Includes companion CD-ROM.

Order number 1-56690-135-9
312 pages, 7" x 9" softcover

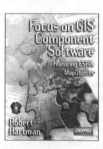

Focus on GIS Component Software: Featuring ESRI's MapObjects

Robert Hartman

This book explains what GIS component technology means for managers and developers. The first half is oriented toward decision makers and technical managers. The second half is oriented toward programmers, illustrated through hands-on tutorials using Visual Basic and ESRI's MapObject product. Includes companion CD-ROM.

Order number 1-56690-136-7
368 pages, 7" x 9" softcover

INSIDE ArcView GIS, Second Edition

Scott Hutchinson and Larry Daniel

Written for the professional seeking quick proficiency with ArcView, this new edition provides tips on making the transition from Release 2 to 3 and an overview of the Spatial Analyst and Network Analyst extensions. The book also presents the software's principal functionality through the development of an application from start to finish, along with several exercises. A companion CD-ROM includes sample application files and exercise data sets, utilities and third party product information.

Order number 1-56690-116-2
500 pages, 7" x 9" softcover

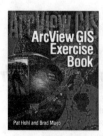

ArcView GIS Exercise Book, Second Edition

Pat Hohl and Brad Mayo

Written to Version 3.x, this book includes exercises on manipulation of views, themes, tables, charts, symbology, layouts and hot links, and real world applications such as generating summary demographic reports and charts for market areas, environmental risk analysis, tracking real estate listings, and customization for task automation.

Order number 1-56690-124-3
480 pages, 7" x 9" softcover

ArcView GIS/Avenue Programmer's Reference, Third Edition

Amir Razavi and Valerie Warwick

This all-new edition of the popular *ArcView GIS/Avenue Programmer's Reference* has been fully updated based on ArcView GIS 3.1. Included is information on more than 200 Avenue classes, plus 101 ready-to-use Avenue scripts—all organized for optimum accessibility. The class hierarchy reference provides a summary of classes, class requests, instance requests, and enumerations. The Avenue scripts enable readers to accomplish a variety of common customization tasks, including manipulation of views, tables, FThemes, IThemes, VTabs, and FTabs; script management; graphical user interface management; and project production documentation.

Order number 1-56690-170-7
544 pages, 7 3/8" x 9 1/8"

ArcView GIS Avenue Scripts: The Disk, Third Edition

Valerie Warwick

All of the scripts from the *ArcView GIS/Avenue Programmer's Reference,Third edition,* with installation notes, ready-to-use on disk. Written to Release 3.1.

Order number 1-56690-171-5
3.5" disk

ArcView GIS/Avenue Developer's Guide, Third Edition

Amir Razavi

This books continues to offer readers one of the most complete introductions to Avenue, the programming language of ArcView GIS. By working through the book, intermediate and advanced ArcView GIS users will learn to customize the ArcView GIS interface; create, edit, and test scripts; produce hardcopy maps; and integrate ArcView GIS with other applications.

Order number 1-56690-167-7

432 pages, 7" x 9" softcover

MapBasic Developer's Guide

Angela Whitener and Breck Ryker

MapBasic is the programming language for the popular desktop mapping software, MapInfo Professional. Written to 4.x, *MapBasic Developer's Guide* is a handbook for custoimizing MapInfo Professional. The book begins with a tutorial on MapBasic elements, the MapBasic development environment, and program building basics. Subsequent chapters focus on customizing and editing of all program components. Companion disk included.

Order number 1-56690-113-8

608 pages, 7" x 9" softcover

INSIDE MapInfo® Professional, Second Edition

Angela Whitener, Paula Loree, and Larry Daniel

Based on Release 5.x

Now updated to the software's latest features and functions, this book continues to set the standard for desktop mapping how-to and reference manuals. Essential software functions are revealed through development of a single application from start to finish. Step-by-step examples, case studies, notes, and tips are also located throughout the book to assist readers in their quest to make optimal use of one of today's most popular desktop mapping applications.

Order number 1-56690-186-3

500 pages, 7" x 9" softcover with CD-ROM

GIS: A Visual Approach

Bruce Davis

This is a comprehensive introduction to the application of GIS concepts. The book's unique layout provides clear, highly intuitive graphics and corresponding concept descriptions on the same or facing pages. It is an ideal general introduction to GIS concepts.

Order number 1-56690-098-0

400 pages, 7" x 9"

GIS: A Visual Approach Graphic Files

This set of 12 disks includes 137 graphic files in Adobe Acrobat, plus the Acrobat Reader. Corresponding with chapters in *GIS: A Visual Approach*, nearly 90% of the book's images are included. Available in Windows or Mac platforms. Ideal for instructors and organizations with a large GIS user base.

Order number 1-56690-120-0

Set of disks